Climbing Mount Improbable

By the Same Author

THE SELFISH GENE

THE EXTENDED PHENOTYPE

THE BLIND WATCHMAKER

RIVER OUT OF EDEN

Original drawings by Lalla Ward

CLIMBING MOUNT IMPROBABLE

Richard Dawkins

W · W · Norton & Company New York London

First American Edition 1996

First published as a Norton paperback 1997

Printed in the United States of America

The text of this book is composed in Centaur. Composition by Justine Burkat Trubey using Adobe Pagemeaker 6.0. Manufacturing by The Haddon Craftsmen, Inc.
Book design by Margaret M. Wagner

Library of Congress Cataloging-in-Publication Data

Dawkins, Richard, 1941–
Climbing mount improbable / Richard Dawkins ; original drawings by Lalla Ward.
p. cm.
Includes bibliographical references and index.
ISBN 0-393-03930-7
1. Natural selection. 2. Evolutionary genetics. 3. Morphogenesis. I. Title.
QH375.D376 1996
575.01'62—dc20
96–19138
CIP

ISBN 0-393-31682-3 pbk.

W. W. Norton & Company, Inc.
500 Fifth Avenue, New York, N.Y. 10110
www.wwnorton.com

W. W. Norton & Company Ltd.
Castle House, 75/76 Wells Street, London W1T 3QT

9 0

For Robert Winston,
a good doctor and a good man

Contents

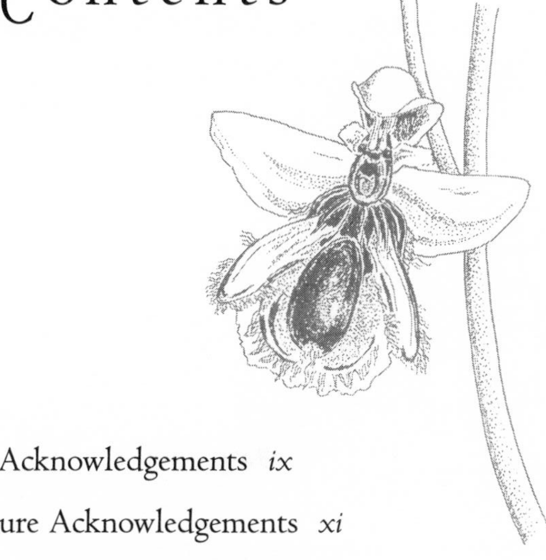

Acknowledgements

THIS BOOK GREW OUT OF MY ROYAL INSTITUTION Christmas Lectures, televised by the BBC under the general title *Growing Up in the Universe.* I have had to abandon that title because at least three other books have since appeared with almost identical names. Moreover, my book itself has grown up and changed, so it is no longer fair to call it the book of the Christmas Lectures. Nevertheless I should like to thank the Director of the Royal Institution for honouring me with the invitation to join the historic lineage of Christmas Lecturers going all the way back to Michael Faraday. Bryson Gore of the Royal Institution, together with William Woollard and Richard Melman of Inca Television, influenced the lectures greatly, and traces of their influence will still be found in this greatly transformed and enlarged book.

Michael Rodgers read and constructively criticized early drafts of more chapters than are here printed, and advised decisively on the reconstruction of the whole book. Fritz Vollrath and Peter Fuchs gave expert readings of Chapter 2, while Michael Land and Dan Nilsson did the same for Chapter 5. All four of these experts gave generously of their knowledge when I tapped it. Mark Ridley, Matt Ridley, Charles Simonyi and Lalla Ward Dawkins read the whole book in a late draft and provided helpful criticism and reassuring encouragement in needful proportions. Mary Cunnane of W. W. Norton and Ravi

Mirchandani of Viking Penguin showed generous tolerance and big-hearted judgement as the book grew, took on a life of its own and finally shrank again to more manageable scope. John Brockman lurked encouragingly in the background, never interfering but always ready with support. Computer experts are heroes, too often unsung. In this book I have used the programs of Peter Fuchs, Thiemo Krink and Sam Zschokke. Ted Kaehler collaborated with me in conceiving and writing the difficult Arthromorphs program. In my own suite of 'watchmaker' programs I have frequently benefited from the advice and help of Alan Grafen and Alun ap Rhisiart. The staff of the Zoological and Entomological Collections of the University Museum at Oxford lent specimens and expert advice. Josine Meijer was a willing and resourceful picture researcher. My wife, Lalla Ward Dawkins, did the drawings (though not the layouts) and her love of Darwinian Creation shines through every one of them.

I should like to thank Charles Simonyi, not only for his immense generosity in endowing the post in Public Understanding of Science which I now hold at Oxford, but also for articulating his vision—which coincides with mine—of the craft of explaining science to a large audience. Do not talk down. Try to inspire everybody with the poetry of science and make your explanations as easy as honesty allows, but at the same time do not neglect the difficult. Put extra effort into explaining to those readers prepared to put matching effort into understanding.

Picture Acknowledgements

Drawings by Lalla Ward: 1.7, 1.9, 1.10, 1.13, 1.14, 2.9, 3.1, 3.3, 4.2, 4.3, 4.4, 4.5, 4.6, 4.7, 5.1, 5.15, 6.3, 6.4, 6.10, 6.13, 6.15, 7.3, 7.8, 7.15a, 7.16, 8.2, 8.3, 8.6; 1.2 (after Hölldobler and Wilson); 1.3 (after Wilson); 1.11 (after Eberhard); 2.6 (after Bristowe); 5.30 (after M. F. Land); 7.10 (after Brusca and Brusca); 7.11 (after *Collins Guide to Insects*); 7.17 (after Brusca and Brusca); 10.6 (after Heijn from Ulenberg).

Computer-generated images by the author: 1.14, 1.15, 1.16, 5.3*, 5.5*, 5.6*, 5.7*, 5.9*, 5.10*, 5.11*, 5.12, 5.20*, 5.28, 6.2*, 6.3*, 6.5, 6.6, 6.8, 6.11, 6.12, 6.14, 7.1, 7.9, 7.12, 7.13, 7.14 (images marked with an asterisk redrawn by Nigel Andrews); by Jeremy Hopes 5.13.

Heather Angel: 1.5, 1.11b, 5.21, 8.1. Ardea: 1.8 (Hans D. Dossenbach), 1.11a (Tony Beamish), 6.7 (P. Morris), 9.3e (Bob Gibbons). Euan N. K. Clarkson: 5.28. Bruce Coleman: 10.3a (Gerald Cubitt). W. D. Hamilton: 10.1, 10.2, 10.4, 10.5, 10.7. Ole Munk: 5.31. NHPA: 6.1 (James Carmichael Jr). Chris O'Toole: 1.6a and b. Oxford Scientific Films: 1.4 (Rudie Kuiter), 2.1 (Densey Clyne), 5.19 (Michael Leach), 5.19b (J. A. L. Cooke), 10.2b (K. Jell), 10.3b (David Cayless). Portech Mobile Robotics Laboratory, Portsmouth: 9.2. Prema Photos: 8.5 (K. G. PrestonMafham). David M. Raup: 6.9. Science Photo Library: 9.3a (A. B. Dowsett), 9.3b (John Bavosi), 9.3c (Manfred Kage), 9.3d (David Patterson), 9.6 (J. C. Revy). Dr Fritz Vollrath: 2.2, 2.3, 2.4, 2.10, 2.11, 2.12, 2.13. Zefa: 9.1.

1.1 from Michell, J. (1978) *Simulacra*. London: Thames and Hudson.

2.5 from Hansell (1984).

2.7 and 2.8 from Robinson (1991).

2.14 and 2.15 from Terzopoulos *et al.* (1995) © 1995 by the Massachusetts Institute of Technology.

3.2 courtesy of the *Hamilton Spectator*, Canada.

4.1 courtesy of J. T. Bonner 1965, © Princeton University Press.

5.2 from Dawkins (1986) (drawing by Bridget Peace).

5.4a, b and d, 5.8a-e, 5.24a and b from Land (1980) (redrawn from Hesse, 1899).

5.4c from Salvini-Plawen and Mayr (1977) (after Hesse, 1899).

5.16a and b Hesse from Untersuchungen¸ber die organe der Lichtempfindung bei niederen thieren, *Zeitschrift für Wissenschaftliche Zoologie*, 1899.

5.17, 5.19d and e, 5.25, 5.26 courtesy of M. F. Land.

5.18a and f, 5.27, 5.30 drawings by Nigel Andrews.

5.22 drawing by Kuno Kirschfeld, reproduced by permission of Naturwissenschaftliche Rundschau, Stuttgart.

5.23 courtesy of Dan E. Nilsson from Stavenga and Hardie (eds.) (1989).

5.29a-e courtesy of Walter J. Gehring *et al.*, from Georg Halder *et al.* (1995).

6.16 from Meinhardt (1995).

7.2, 7.4, 7.5, 7.6, 7.7 from Ernst Haeckel (1904) *Kunstformen der Natur*. Leipzig and Vienna: Verlag des Bibliographischen Instituts.

7.15b from Raff and Kaufman (1983) (after Y. Tanaka, 'Genetics of the Silkworm', in *Advances in Genetics* 5: 239–317, 1953).

8.4 from Wilson (1971) (from Wheeler, 1910, after F. Dahl).

9.4 Jean Dawkins.

9.5 © K. Eric Drexler, Chris Peterson and Gayle Pergamit. All rights reserved. Reprinted with permission from *Unbounding the Future: The Nanotechnology Revolution*. William Morrow, 1991.

Climbing Mount Improbable

CHAPTER I

FACING MOUNT RUSHMORE

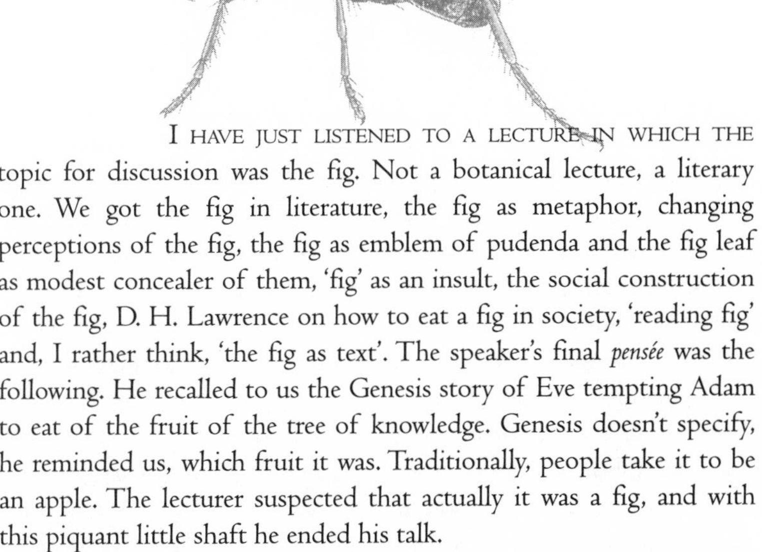

I HAVE JUST LISTENED TO A LECTURE IN WHICH THE topic for discussion was the fig. Not a botanical lecture, a literary one. We got the fig in literature, the fig as metaphor, changing perceptions of the fig, the fig as emblem of pudenda and the fig leaf as modest concealer of them, 'fig' as an insult, the social construction of the fig, D. H. Lawrence on how to eat a fig in society, 'reading fig' and, I rather think, 'the fig as text'. The speaker's final *pensée* was the following. He recalled to us the Genesis story of Eve tempting Adam to eat of the fruit of the tree of knowledge. Genesis doesn't specify, he reminded us, which fruit it was. Traditionally, people take it to be an apple. The lecturer suspected that actually it was a fig, and with this piquant little shaft he ended his talk.

This kind of thing is the stock-in-trade of a certain kind of literary mind, but it provokes *me* to literal-mindedness. The speaker obviously knew that there never was a Garden of Eden, never a tree of knowledge of good and evil. So what was he actually trying to say? I suppose he had a vague feeling that 'somehow', 'if you will', 'at some level', 'in some sense', 'if I may put it this way' it is somehow 'right' that the fruit in the story 'should' have been a fig. But enough of this. It is not that we should be literalist and Gradgrindian, but our elegant lecturer was *missing* so much. There is genuine paradox and real poetry lurking in the fig, with subtleties to exercise an inquiring mind and wonders to uplift an aesthetic one. In this book I want to move to a position where I can tell the true story of the fig. But the fig story is only one out of millions that all have the same Darwinian grammar and logic—albeit the fig story is among the most satisfyingly intricate in evolution. To anticipate the central metaphor of the book, the fig tree stands atop one of the highest peaks on the massif of Mount Improbable. But peaks as high as the fig's are best conquered at the end of the expedition. Before that there is much that needs to be said, a whole vision of life that needs to be developed and explained, puzzles that must be solved and paradoxes that must be disarmed.

As I said, the story of the fig is, at the deepest level, the same story as for every other living creature on this planet. Though they differ in surface detail, all are variations on the theme of DNA and the 30 million ways by which it propagates itself. On our route we shall have occasion to look at spider webs—at the bewildering, though unconscious, ingenuity with which they are made and how they work. We shall reconstruct the slow, gradual evolution of wings and of elephant trunks. We shall see that 'the' eye, legendarily difficult though its evolution sometimes seems, has actually evolved at least forty and probably sixty times independently all around the animal kingdom. We shall program computers to assist our imagination in moving easily through a gigantic museum of all the countless creatures that have ever lived and died, and their even more numerous imaginary cousins, who have never been born. We shall wander the paths of Mount Improbable, admiring its vertical precipices from afar, but always restlessly seeking the gently graded slopes on the other side. The meaning of the parable of Mount Improbable will be made clear, and much else besides. I need to begin by clarifying the problem of apparent design in nature, its relationship to true, human design and its relationship to chance. This is the purpose of Chapter 1.

The Natural History Museum in London has a quirky collection of stones that chance to resemble familiar objects: a boot, a hand, a baby's skull, a duck, a fish. They were sent in by people who genuinely suspected that the resemblance might mean something. But ordinary stones weather into such a welter of shapes, it is not surprising if occasionally we find one that calls to mind a boot, or a duck. Out of all the stones that people notice as they walk about, the museum has preserved the ones that they pick up and keep as curiosities. Thousands of stones remain uncollected because they are just stones. The coincidences of resemblance in this museum collection are meaningless, though amusing. The same is true when we think we see faces, or animal shapes, in clouds or cliff profiles. The resemblances are accidents.

The craggy hillside in Figure I.I is supposed to suggest the profile of the late President Kennedy. Once you have been told, you can just see a slight resemblance to either John or Robert Kennedy. But some don't see it and it is certainly easy to believe that the resemblance is accidental. You couldn't, on the other hand, persuade a reasonable person that Mount Rushmore, in South Dakota, had just happened to weather into the features of Presidents Washington, Jefferson, Lincoln and Theodore Roosevelt. We do not need to be told that these were deliberately carved (under the direction of Gutzon Borglum). They are obviously not accidental: they have design written all over them.

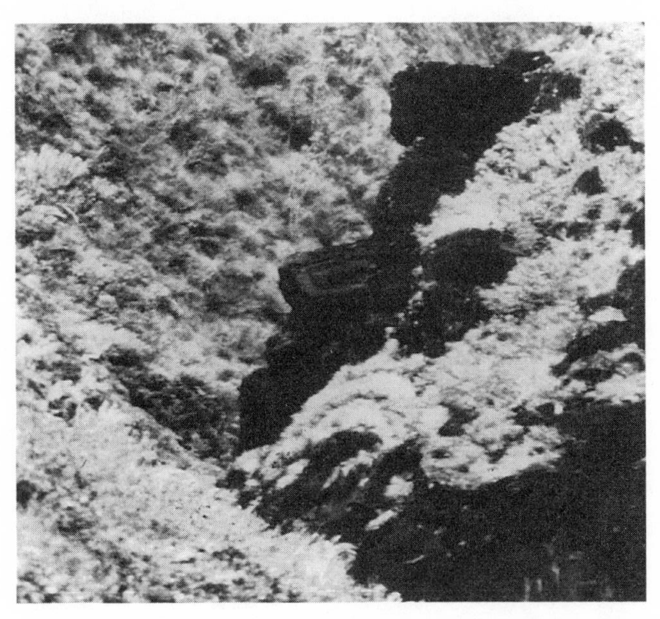

Figure I.I A pure accident. President Kennedy's profile in a hillside in Hawaii.

The difference between Mount Rushmore and the weathered likeness of John Kennedy (or Mont St Pierre in Mauritius or other such curiosities of natural weathering) is this. The sheer number of details in which the Mount Rushmore faces resemble the real things is too great to have come about by chance. The faces are clearly recognizable, moreover, when seen from different angles. Figure 1.1's chance resemblance to President Kennedy, on the other hand, is noticed only if the cliff is seen from a particular angle and in a particular light. Yes, a rock can weather into the shape of a nose seen from a certain vantage point, and maybe a couple of other rocks happen to have tumbled into the shape of lips. It is not much to ask of chance that it should produce a modest coincidence like this, especially if the photographer has every possible angle to choose from and only one gives the resemblance (and there is the added fact, which I'll return to in a moment, that the human brain seems actively eager to see faces: it seeks them out). But Mount Rushmore is another matter. Its four heads are clearly *designed.* A sculptor conceived them, drew them out on paper, made meticulous measurements all over the cliff, and supervised teams of workmen who wielded pneumatic drills and dynamite to carve out the four faces, each sixty feet high. The weather *could* have done the same job as the artfully deployed dynamite. But of all the possible ways of weathering a mountain, only a tiny minority would be speaking likenesses of four particular human beings. Even if we didn't know the history of Mount Rushmore, we'd estimate the odds against its four heads being carved by accidental weathering as astronomically high—like tossing a coin forty times and getting heads every time.

I think that the distinction between accident and design is clear, in principle if not always in practice, but this chapter will introduce a third category of objects which is harder to distinguish. I shall call them *designoid* (pronounced 'design-oid' not 'dezziggnoid'). Designoid objects are living bodies and their products. Designoid objects *look* designed, so much so that some people—probably, alas, most people—think that they *are* designed. These people are wrong. But they are right in their conviction that designoid objects cannot be the result of chance. Designoid objects are not accidental. They have in fact been

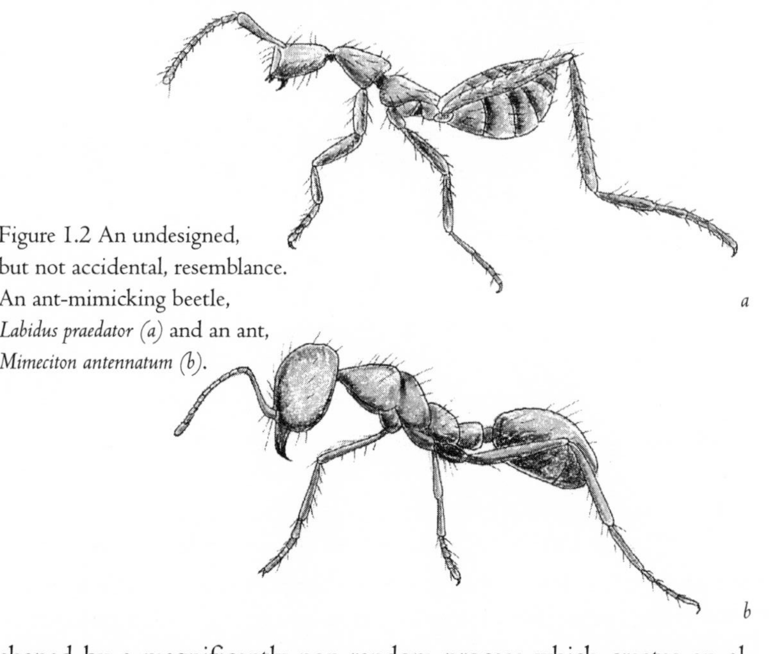

Figure I.2 An undesigned, but not accidental, resemblance. An ant-mimicking beetle, *Labidus praedator* (*a*) and an ant, *Mimeciton antennatum* (*b*).

shaped by a magnificently non-random process which creates an almost perfect illusion of design.

Figure I.2 shows a living sculpture. Beetles in general don't look like ants. So, if I see a beetle that looks almost exactly like an ant—a beetle, moreover, that makes its living entirely in an ants' nest—I shall rightly suspect that the coincidence means something. The top animal is actually a beetle—its closer cousins are common or garden beetles—but it looks like an ant, walks like an ant, and lives among ants in an ants' nest. The one at the bottom is a real ant. As with any realistic statue, the resemblance to the model is not an accident. It demands an explanation other than sheer chance. What kind of an explanation? Since all beetles that look strikingly like ants live in ants' nests, or at least in close association with ants, could it be some chemical substance from the ants, or some infection from the ants, rubbing off on the beetles and changing the way they grow? No, the true explanation—Darwinian natural selection—is very different, and we shall come to it later. For the moment, it is enough that we are sure this resemblance, and other examples of 'mimicry', are not acci-

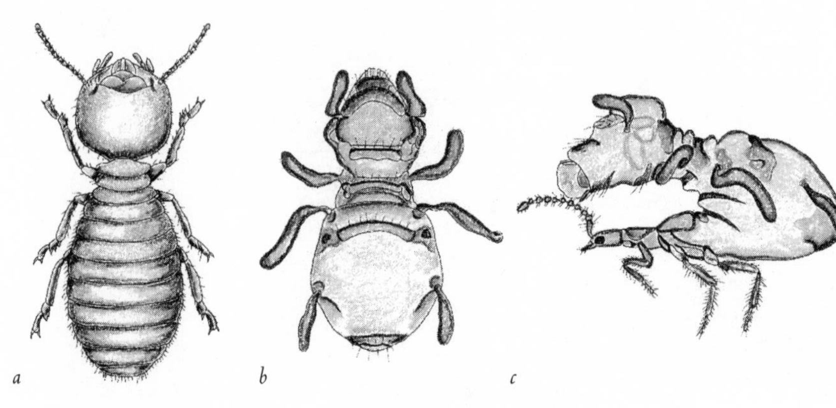

Figure I.3 *(a)* A real termite, *Amitermes hastatus; (b)* a beetle, *Coatonachthodes ovambolandicus,* mimicking a termite; *(c)* how the trick is done.

dental. They are either designed or they are due to some process that produces results just as impressive as design. We shall look at some other examples of animal mimicry, leaving open, for the moment, the explanation of how these remarkable resemblances come about.

The previous example shows what a good job beetle flesh can do if it 'sets out to mimic' a different kind of insect. But now look at the creature in Figure I.3b. It appears to be a termite. Figure I.3a is a real termite, for comparison. The specimen in Figure I.3b is an insect, but it is not a termite. It is, in fact, a beetle. I admit that I've seen better mimics in the insect world, including the ant-mimicking beetle of the previous example. The 'beetle' here is just a little odd. Its legs seem to lack proper joints, like little twisty balloons. Since, like any other insect, a beetle has jointed legs at its disposal, you might hope for a better shot at mimicking a termite's jointed legs. So, what is the solution to this conundrum? Why does this mimic look like an inflated dummy rather than like a real, jointed insect? The answer can be seen in Figure I.3c, which is one of the most astonishing spectacles in all natural history. It shows the termite-mimicking beetle in side view. The true head of the beetle is a diminutive affair (you can see the eye just near the normal, jointed antennae), attached to a slender trunk or thorax bearing three normal, jointed beetle legs, on which it actually walks. It is with the abdomen that the trick is done. It is arched backwards so that it hangs over and completely covers the head, thorax and legs like a parasol. The entire 'termite' is constructed from the (anatomically) rear half of the beetle's abdomen. The 'termite head' is the rear tip of the beetle's abdomen. The 'termite legs' and 'antennae' are flapping excrescences of the abdomen. No wonder the quality of the mimicry is not quite up to the standard of the beetle's ant-mimicking cousin of the previous picture. This termite-mimicking beetle, by the way, lives in termite nests, making its living as a parasite in much the same way as Figure I.2's ant-mimicking beetle makes its living among ants. Although the quality of the resemblance is less, when you consider its starting materials the termite-mimicking beetle seems to achieve a more impressive feat of sculpture than the ant mimicking beetle. This is because the ant-mimic does it by modifying each bit of its body to look like the corresponding bit of the ant's body. But the termite-mimic does it by modifying a completely different bit of itself—the abdomen—to look like all the bits of the termite.

My own favourite among animal 'statues' is the leafy sea dragon (Figure I.4). It is a fish, a kind of sea-horse, whose body is sculpted into the shape of seaweed. This gives it protection, for it lives among seaweed and is remarkably difficult to see there. Its mimicry is too uncannily good to

Figure I.4 Perfection of camouflage. A female leafy sea dragon, *Phycodurus eques*, from Australia.

be accidental in any simple sense. It lies closer to Mount Rushmore than to the Kennedy cliff. My confidence is based partly upon the sheer number of ways in which it impresses us by looking like something that it isn't; and partly upon the fact that fish don't normally have projections of anything like that shape. In this respect the leafy sea dragon's feat compares with the termite-mimic, rather than the ant-mimic.

So far we have talked of objects that impress us as realistic sculptures do, objects that we feel can't be accidental because they look too strikingly like other objects. Leafy sea dragons and ant-mimicking beetles are designoid statues: they overwhelmingly look as if they have been designed by an artist to resemble something else. But statues are only one kind of object that humans design. Other human artefacts impress us not by resembling something but by being unmistakably useful for some purpose. An aeroplane is useful for flying. A pot is useful for holding water. A knife is useful for cutting things.

If you offered a reward for stones that were naturally sharp enough to cut things, and also for stones that happened to be of a shape to hold water, you'd probably be sent some effective makeshifts. Flints often fracture in such a way as to leave a good keen edge, and if you wandered the quarries and screes of the world you'd certainly find some handy natural blades. Among the richness of shapes into which stones can weather, some would happen to include concavities that hold water. Certain types of crystal naturally encrust around a hollow, albeit chunky, sphere which, when it splits in half, yields two serviceable cups. These stones even have a name: geode. I use a geode as a paperweight on my desk, and I'd use it to drink from if its interior were not roughly pitted and therefore hard to wash.

It is easy to devise measures of efficiency that would show up natural pots as less efficient than man-made ones. Efficiency is some measure of benefit divided by cost. The benefit of a pot could be measured as the quantity of water that it holds. Cost can conveniently be measured in equivalent units: the quantity of the material of the pot itself. Efficiency might be defined as the volume of water that a pot can hold divided by the volume of material that goes to make the pot itself. The hollow stone on my desk holds 87.5 cc of water. The volume of the stone itself (which I measured by Archimedes' famous

Eureka-in-the-Bath method) is 130 cc. The efficiency of this 'pot' is therefore about 0.673. This is a very low efficiency, not surprisingly so since the stone was never designed to hold water. It just happens to hold water. I have just done the same measurements on a wineglass, whose efficiency turns out to be about 3.5. A friend's silver cream jug is even more efficient. It holds 250 cc of water while the silver of which it is made displaces a mere 20 cc. Its efficiency is therefore as high as 12.5.

Not all human-designed pots are efficient in this sense. A chunky pot from the kitchen cupboard holds 190 cc of water while using up a massive 400 cc of marble. Its 'efficiency' is therefore only 0.475, even lower than the totally undesigned hollow stone. How can this be? The answer is revealing. This marble pot is in fact a mortar. It is not designed to hold liquid. It is a hand mill for grinding spices and other foods with a pestle: a stout rod which is wielded with great force against the inside of the mortar. You couldn't use a wineglass as a mortar: it would shatter under the force. The measure of efficiency that we devised for pots is not suitable when the pot is designed as a mortar. We should devise some other benefit/cost ratio, where benefit takes account of strength against being broken by a pestle. Would the natural geode, then, qualify as a well-designed mortar? It would probably pass the strength test but if you tried to use it as a mortar its rough and craggy interior would soon prove a disadvantage, the crevices protecting grains from the pestle. You'd have to improve your measure of the efficiency of a mortar by including some index of smoothness of internal curvature. That my marble mortar is designed can be discerned from other evidence: its perfectly circular plan section, coupled with its elegantly turned lip and plinth seen in elevation.

We could devise similar measures of the efficiency of knives, and I have no doubt that the naturally flaked flints that we happen to pick up in a quarry would compare unfavourably, not only with Sheffield steel blades but with the elegantly sculpted flints that museums display in Late Stone Age collections.

There is another sense in which natural, accidental, pots and knives are inefficient compared with their designed equivalents. In the course of finding one usefully sharp flint tool, or one usefully watertight stone vessel, a huge number of useless stones had to be examined and discarded. When we measure the water held by a pot, and divide by the volume of stone or clay in the material of the pot, it might be fairer to add into the denominator the cost of the stone or clay discarded. In the case of a man- made pot thrown on a wheel, this additional cost would be negligible. In the case of a carved sculpture the cost of discarded chippings would be present but small. In the case of the accidental, *objet trouvé* pot or knife, the 'discard cost' would be colossal. Most stones don't hold water and are not sharp. An industry that was entirely based upon *objets trouvés,* upon *found* objects as tools and utensils, rather than artificially shaped tools and utensils, would have a huge dead weight of inefficiency in the spoil heaps of alternatives discarded as useless. Design is efficient compared with finding.

Let's turn our attention now to designoid objects—living things that look as though they have been designed but have actually been put together by a completely different process—beginning with designoid pots. The pitcher plant (Figure 1.5) could be seen as just another kind of pot, but it has an elegant 'economy ratio', comparable to the wineglass that I measured, if not the silver jug. It gives every appearance of being excellently well designed, not just to hold water but to drown insects and digest them. It concocts a subtle perfume which insects find irresistible. The smell, abetted by a seductive colour pattern, lures prey to the top of the pitcher. There the insects find themselves on a steep slide whose treacherous slipperiness is more than accidental, set about with downward-facing hairs well placed to impede their last struggle. When they fall, as they nearly always do, into the dark belly of the pitcher, they find more than just water in which to drown. The details, brought to my attention by my colleague Dr Barrie Juniper, are remarkable and I'll briefly tell the story.

It is one thing to trap insects, but the pitcher plant lacks jaws, muscles and teeth with which to reduce them to a state fit for digesting. Perhaps plants could grow teeth and munching jaws, but in practice there is an easier solution. The water in the pitcher is home to a rich community of maggots and other creatures. They live no-

Figure I.5 A designoid pot. Pitcher plant, *Nepenthes pervillei,* from the Seychelles.

where else but in the enclosed ponds created by pitcher plants, and they are endowed with the jaws that the plant itself lacks. The corpses of the pitcher plant's drowned victims are devoured and decomposed by the mouthparts and digestive juices of its maggot accomplices. The plant itself subsists on the detritus and excretory products, which it absorbs through the lining of the pitcher.

The pitcher plant doesn't just passively accept the services of maggots that happen to fall into its private pool. The plant works actively to provide the maggots with a service that they need in their turn. Analyse the water in a pitcher plant and you find a singular fact. It is not fetid, as might be expected of standing water in such conditions, but strangely rich in oxygen. Without this oxygen the vital maggots could not flourish, but where does it come from? It is manufactured by the pitcher plant itself, and the plant gives every apparent indication of being specifically designed to oxygenate the water. The cells that line the pitcher are richer in oxygen-producing chlorophyll than the outside cells that face the sun and air. This surprising reversal of apparent common sense is explicable: the inside cells are specialized to secrete oxygen directly into the water inside the pitcher. The pitcher plant does not just borrow its vicarious jaws: it hires them, paying in the currency of oxygen.

Other designoid traps are common. The Venus's fly-trap is as elegant as the pitcher plant, with the added refinement of moving parts. The insect prey releases the trap by triggering sensitive hairs on the plant, whose jaws smartly close. The spider web is the most familiar of all animal traps, and we shall do it justice in the next chapter. An underwater equivalent is the net constructed by stream-dwelling caddis fly larvae. Caddis larvae are also notable for their feats as builders of houses for themselves. Different species use stones, sticks, leaves or tiny snail shells.

A familiar sight in various parts of the world is the conical trap of the ant-lion. This fearsome creature is the larva of—what could sound more gentle?—a lacewing fly. The ant-lion lurks just under the sand at the bottom of its pit, waiting for ants or other insects to fall in. The pit achieves its almost perfectly conical shape—which makes it hard for victims to claw their way out—not by design but as a consequence of some simple rules of physics, exploited by the way the ant-lion digs. From the bottom of the descending pit, it flicks sand right over the edge with a jerk of the head. Flicking sand from the bottom of a pit has the same effect as draining an hourglass from below: the sand forms itself naturally into a perfect cone of predictable steepness.

Figure I.6 brings us back to pots. Many solitary wasps lay their eggs on prey, which have been stung to paralysis and then hidden in a hole. They seal the hole up so that it is invisible, the larva feeds on the prey inside and finally emerges as a winged adult to complete the

a

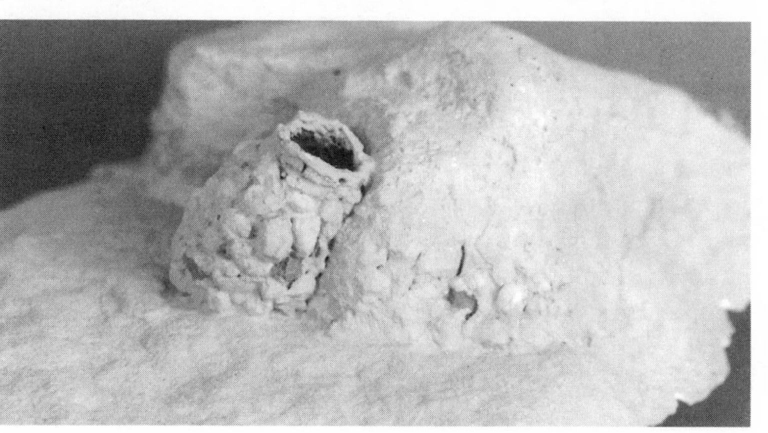

b

Figure I.6 Designoid pots made by animal artisans: *(a)* potter wasp and *(b)* mason bee.

cycle. Most species of solitary wasp dig their nesting hole in the ground. The potter wasp makes its 'hole' out of clay—a round pot, up a tree, mounted inconspicuously on a twig (Figure I.6a). Like the pitcher plant, this pot would score favourably on our efficiency test for apparent design. Solitary bees show a similar pattern of nesting in holes, but they feed their larvae on pollen instead of animal prey. Like the potter among wasps, many species of mason bee build their own pot nest. The pot in Figure I.6b is made not from clay but from small stones cemented together. Apart from its resemblance to an efficient, man-made receptacle, there is something else rather wonderful about the particular specimen photographed. You see only one pot here, but there are actually four. The other three have been covered by the bee with hardened mud, to give an exquisite match to the surrounding rock. No predator would ever find the young larvae growing up in the pots. The only reason this cluster was seen, by my colleague Christopher O'Toole on a visit to Israel, is that the bee had not quite finished covering the last pot.

These insect pots have all the hallmarks of 'design'. In this case, unlike the pitcher plant, they really were fashioned by the actions of a skilled—albeit probably unconsciously so—creature. The pots of the potter wasp and mason bee seem, on the face of it, closer to man-made pots than to the pitcher plant. But the wasp and the bee didn't consciously or deliberately design their pots. Although they were shaped, out of clay or stones, by behavioural actions of the insects, this is not importantly different from the way the insects' own bodies were made during embryonic development. This may sound odd but let me explain. The nervous system grows in such a way that the muscles and limbs and jaws of the living wasp move in certain coordinated patterns. The consequence of these particular clockwork limb movements is that clay is gathered and fashioned into the shape of a pot. The insect very probably does not know what it is doing, nor why it is doing it. It has no concept of a pot as a work of art, or as a container, or as a brood chamber. Its muscles just move in the way its nerves dictate, and a pot is the result. So for this reason we firmly—if wonderingly—classify the wasp and bee pots as designoid not designed: not shaped by the animal's own creative volition. Actually, to

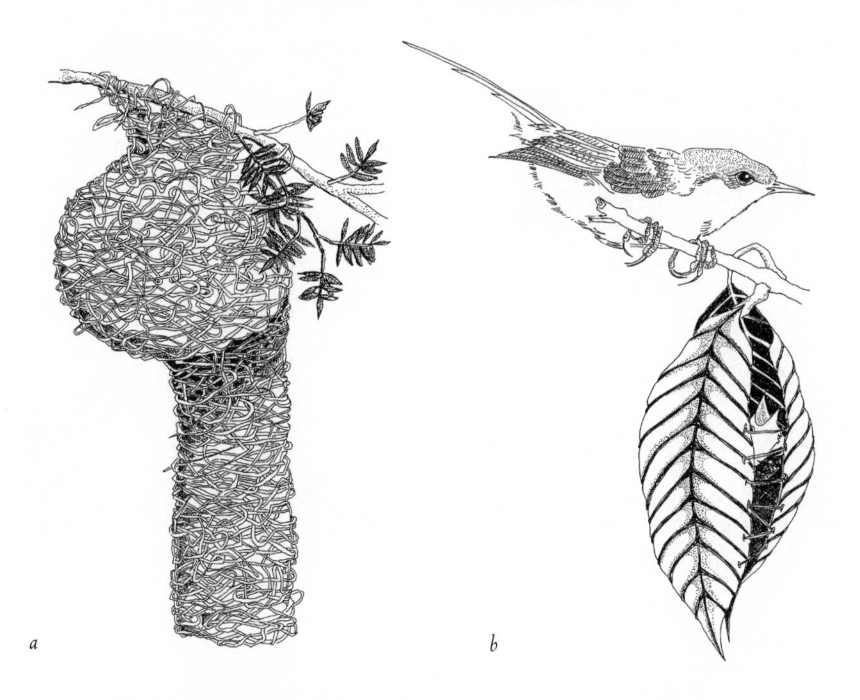

Figure I.7 Designoid craftsmanship. (*a*) Weaverbird nest and (*b*) tailorbird, *Orthotomus sutorius*, with its nest.

be fair, I cannot know for certain that wasps lack creative volition and true design. It is enough for me that my explanation works even if they do. The same goes for birds' nests (Figure I.7) and bowers, caddis houses and caddis nets, but not for the sculptures of Mount Rushmore or the tools used to hew them—they really are designed.

Karl von Frisch, the famous Austrian zoologist who deciphered the bee dance, once wrote: 'If we imagined for a moment that termites were as tall as human beings, their tallest hillocks, enlarged on the same scale, would be nearly a mile high, four times the height of New York's Empire State Building.' The skyscrapers of Figure I.8 were made by Australian compass termites. They are called compass termites because their mounds are always lined up north–south—they can be used as compasses by lost travellers (as can satellite dishes, by the way: in Britain they seem all to face south). The advantage of this for the termites is that the broad,

Figure I.8 Insect skyscrapers aligned north to south. Compass termite mounds in Australia.

flat surfaces of the mound are warmed by the early-morning and late-afternoon sun. But the mound is protected from the fierce noonday sun since only the sharp edge is presented to the north—where the sun is at midday in the southern hemisphere. We could be forgiven for thinking the termites had designed this clever trick themselves. But the principle by which their building behaviour appears intelligent is identical to the principle by which the jaws and legs of the termites appear designed. Neither of them is designed. Both are designoid.

Animal artefacts, like caddis and termite houses, birds' nests or mason bee pots, are fascinating, but they are a special case among designoid things—an intriguing curiosity. The name 'designoid' primarily refers to living bodies themselves and their parts. Living bodies are put together not by skilled hands, beaks or jaws, but by the convoluted processes of embryonic growth. A mind addicted to exhaustive classification systems might see artefacts like wasp pots as 'second order designoid objects' or as an intermediate category between designed and designoid, but I think this is simply confusing. Admittedly the pot is made of mud, not living cells, and it is shaped by limb movements which superficially resemble the hand movements of a human potter. But all the 'design', all the elegance, all the fitness of the pot to perform a useful task, come from very different sources in the two cases. The human pot is conceived and planned by a creative process of imagination in the head of the potter, or by deliberate imitation of the style of another potter. The wasp pot gets its elegance and fitness to its task from a very different process—from exactly the same process, indeed, as gave elegance and fitness to the wasp's own body. This will become clearer if we continue our discussion of living bodies as designoid objects.

One of the ways we recognize both true design and designoid pseudo- design is that we are impressed by resemblances between objects and other objects. The Mount Rushmore heads are obviously designed because they resemble real presidents. The leafy sea dragon's resemblance to seaweed is equally obviously not an accident. But mimicry like this, and like the resemblance of beetle to termite, or stick insect to twig, is by no means the only kind of resemblance that impresses us in the living world. Often we are stunned by the resemblance between a living structure and a man-made device that does the same job. The 'mimicry' between human eye and man-made camera is too well known to need illustrating here. Engineers are often the people best qualified to analyse how animal and plant bodies work, because efficient mechanisms have to obey the same principles whether they are designed or designoid.

Often living bodies have converged upon the same shape as each other, not because they are mimicking each other but because the shape that they share is separately useful to each of them. The hedgehog and the spiny tenrec in Figure I.9 are so similar that it seemed almost a waste of effort to draw both of them. They are moderately closely related to each other, both being members of the order Insectivora. Yet other evidence shows that they are sufficiently unrelated for us to be sure that they evolved their prickly appearance independently, presumably for parallel reasons: prickles give protection from predators. Each of the prickly animals is pictured next door to a shrew-like animal which is a closer cousin to it than the other prickly animal is. Figure I.10 gives another example. Animals that swim fast

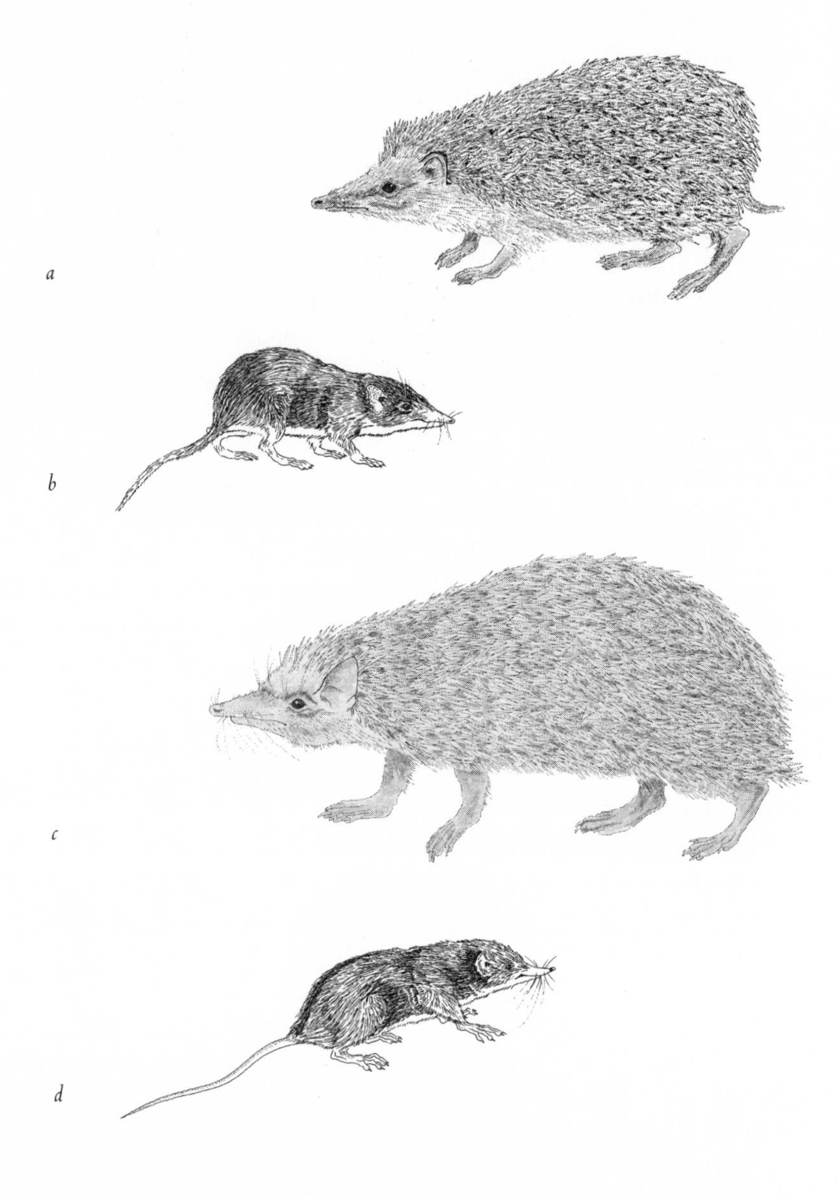

Figure I.9 Animals with similar needs often resemble each other more than they resemble their closer relatives. The Algerian hedgehog, *Erinaceus algirus* (*a*), is a close cousin of the shrew hedgehog, *Neotetracus sinensis* (*b*). The greater hedgehog tenrec, *Setifer setosus* (*c*), is a close cousin of the long-tailed tenrec, *Microgale melanorrachis* (*d*).

Figure 1.10 Convergent evolution: independently evolved streamlining: *(a)* Bottlenose dolphin, *Tursiops truncatus*; *(b)* *Ichthyosaurus*; *(c)* blue marlin, *Makaira nigricans*; and *(d)* Galapagos penguin, *Spheniscus mendiculos*.

near the surface of the sea often converge on the same shape. It is the shape that engineers would recognize as streamlined. The picture shows a dolphin (mammal), an extinct ichthyosaur (which we can think of as the reptilian equivalent of a dolphin), a marlin (bony fish) and a penguin (bird). This kind of thing is called convergent evolution.

Apparent convergence is not always so meaningful. Those people—not all of them missionaries—that dignify face-to-face copulation as diagnostic of higher humanity may be charmed by the millipedes in Figure I.11. If we call this convergence, it is probably not due to convergent needs: rather, there are only so many ways in which a male and a female can juxtapose their bodies, and there could be lots of reasons for hitting upon any one of them.

This brings us full circle to our opening topic of pure accident. There are some living things that resemble other objects but where the resemblance is probably not strong enough to be anything other than accidental. The bleeding heart pigeon has a tuft of red feathers, so placed as to create the illusion of a mortal wound to the breast, but the resemblance is unlikely to mean anything. Equally accidental is the coco-de-mer's resemblance to a woman's loins (Figure I.12a).

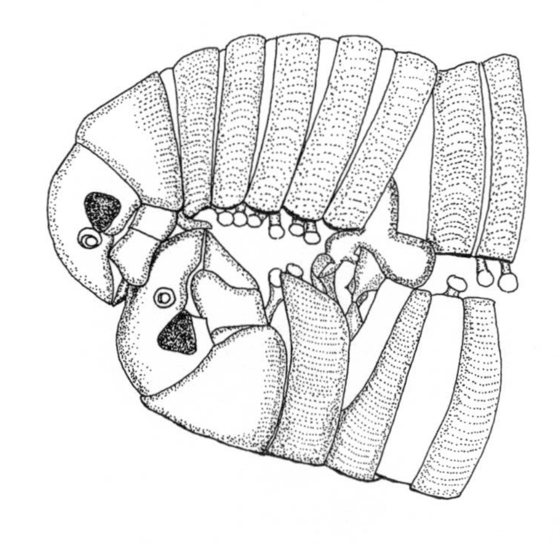

Figure I.11 Copulating millipedes, *Cylindroilus punctatus*, in missionary position.

Figure I.12*a* Accidental resemblances in nature: coco-de-mer.

As in the case of the Kennedy profile in the cliff, the reason for feeling that these resemblances are coincidence is statistical. The pigeon's bleeding heart amounts only to a gash of red feathers. The coco-de-mer's apparent 'mimicry' is admittedly impressive. It involves two or three features, not just one. It even has a suggestion of pubic hair. But the human brain works hard and actively seeks resemblances, especially to parts of our bodies that we find particularly interesting. I suspect that this is going on in our perception of the coco-de-mer, just as it is in our recognition of the Kennedy hillside.

The same goes for the death's-head hawk moth (Figure I.12b). Indeed our brain has an almost indecent eagerness to see faces, which is the basis of one of the most striking illusions known to psychologists. If you get an ordinary face mask from a fancy dress shop and hold it up, hollow side facing another person (with a background that makes the eye holes conspicuous), the viewer is likely to see it stand-

Figure I.12*b* Accidental resemblances in nature: death's-head hawk moth, *Acherontia atropos.*

ing out as a solid face. This has a very odd consequence which you'll discover if you gently rotate the mask from side to side. Remember that the viewer's brain 'thinks' it is a solid face, but the object is actually a hollow mask. When the hollow mask moves to the left, the only way to reconcile the eyes' report with the brain's presumption that the face is solid is to suppose that it moves in the opposite direction. And this is exactly the illusion that the viewer will see. The face will appear to be rotating in a direction opposite to the actual direction of rotation.

So it seems quite likely that the death's-head hawk moth resembles a face by accident. I should add, however, that one of our most respected evolutionary theorists, Robert Trivers, now of Rutgers University, New Jersey, believes that face mimicry on the backs of insects can be an adaptation to scare would-be predators such as birds (we think of the face on the moth as a human skull, but it could equally be a monkey's face). He could well be right, in which case I should have placed the example under my 'designoid' heading. For a different reason, the same may be true of another apparent face mimic, the Japanese samurai crab. This crab has on its back a likeness (not, I have to say, a stunningly impressive likeness) of the fierce features of a samurai warrior. The suggestion has been made that over the centuries Japanese fishermen, abetted by the human brain's natural eagerness to see faces, have noticed a slight resemblance to a face on the backs of some individual crabs. For reasons of superstition or respect, fishermen did not wish to kill crabs with a human-like face (maybe especially a samurai-like face), so they threw them back into the sea instead. Many a crab's life was saved, according to this theory, by its humanoid face, and those crabs with the most pronounced human features in any one generation contributed a disproportionate share of offspring to the next generation. Later generations therefore had a head start over earlier ones, and the resemblance gradually increased.

When we were discussing how to acquire a stone knife simply by finding it, we agreed that you could 'make' a sharp knife by examining all the stones in the world and discarding the blunt ones—the great majority. If you searched enough screes and quarries, you'd surely find a stone with not only a keen blade but a convenient handle as well. It is only a partial oversimplification to say that the pharmaceutical industry works by examining lots of molecules generated at hazard and then testing the efficacy of the minority that seem promising. But we agreed that *finding*, as a method of acquiring a useful implement, was supremely inefficient. Far better to take a suitable material like stone or steel and hone it or sculpt it by design. Yet this is not how designoid objects—living things bearing the illusion of design—are made. Living things come about ultimately by a process that is rather more like 'finding', but it is different from pure finding in a very significant respect.

It may seem an odd fact to notice of a stone, but I'll mention it anyway and follow where it leads. A stone does not have children. If stones had children like themselves, those children would inherit from their parents the attribute of having children. This implies grandchildren and greatgrandchildren for unspecified generations. A far-fetched speculation it might be thought and, in any case, so what? To answer this, turn to something whose sharpness may be equally inadvertent but which does have offspring.

The hard, strap-like leaves of some reeds have quite sharp edges. This sharpness is probably an incidental by-product of other properties of the leaf. You can cut yourself on a reed, enough to annoy but not enough for the sharpness to provoke suspicion of design. No doubt some leaves are sharper than others and you could search the lakeshore for the sharpest reed you could find. Now here is where we part company with stones. Don't just cut with your reed knife, *breed from it.* Or breed from the same plant from which you plucked it. Allow the sharpest plants to cross-pollinate, kill the blunt plants: it doesn't matter how you do it, just see to it that the sharpest plants do most of the breeding. Not once, but generation after generation. As the generations pass you'll notice that there are still bluntish reeds and sharpish reeds around, but the average reed will become steadily sharper. After 100 generations you'd probably have bred something that would give you a decent close shave. If you bred for rigidity at the same time as breeding for keenness of edge, you could eventually cut your throat with a broken reed.

In a sense you have done no more than *find* the quality you seek: no carving, whittling, moulding or whetstone grinding, just finding the best of what is already there. Sharp leaves have been found, blunt leaves have been discarded. It is like the story of finding sharp stones but with one significant addition: the process is cumulative. Stones don't breed whereas leaves, or rather the plants that make leaves, do. Having found the best blade of a generation you don't simply use it until it wears out. You ratchet your gain by breeding from it, transmitting its virtue to the future where it can be built upon. This process is cumulative and never-ending. You are still only finding and finding, but because genetics enables cumulative gain the best specimen you can find in a late generation is better than the best you can find in an earlier generation. This, as we shall see in Chapter 3, is what Climbing Mount Improbable means.

The steadily sharpening reed was an invention to make a point. There are, of course, real examples of the same principle at work. All the plants in Figure 1.13 are derived from one wild species, the wild

Figure 1.13 All these vegetables have been bred from the same ancestor, the wild cabbage, *Brassica olearacea*: (clockwise from top left) Brussels sprout, kohlrabi, Swedish turnip, drumhead cabbage, cauliflower and golden savoy.

cabbage, *Brassica oleracea.* It is a rather nondescript plant which doesn't look much like a cabbage. Humans have taken this wild plant and, over a short period of centuries, shaped it into these really very different kinds of food plants. It is a similar story with dogs (Figure 1.14).

Although hybrids between dogs and jackals and between dogs and coyotes do occur, it is now accepted by most authorities that all breeds of domestic dogs are descended from a wolf ancestor (top left) who lived maybe a few thousand years ago. It is as though we humans had taken wolf flesh and shaped it like a clay pot. But we didn't, of course, literally knead and press wolf flesh into the shape of, say, a whippet or a dachshund. We did it by cumulative finding, or, as it is more conventionally put, selective breeding or artificial selection. Whippet-breeders *found* individuals that looked a little bit more whippet-like than the average. They bred from them, and then found the most whippet-like individuals of the next generation, and so on. Of course it wouldn't have been so relentlessly simple as that, and the breeders wouldn't have had the concept of a modern whippet in their heads as a distant target. Perhaps they just liked the look of the physical characteristics that we would now recognize as whippet-like, or perhaps those visible qualities came along as a by-product of breeding for something else, like proficiency in hunting rabbits. But whippets and dachshunds, Great Danes and bulldogs, were made by a process that resembled finding more than it resembled clay-modelling. Yet it still is not the same as pure finding, because it is cumulative over generations. That is why I call it *cumulative finding.*

Accidental objects are simply found. Designed objects aren't found at all, they are shaped, moulded, kneaded, assembled, put together, carved: in one way or another the individual object is pushed into shape. Designoid objects are cumulatively found, either by humans as in the case of domestic dogs and cabbages, or by nature in the case of, say, sharks. The fact of heredity sees to it that the *accidental* improvements *found* in each generation are accumulated over many generations. At the end of many generations of cumulative finding, a designoid object is produced which may make us gasp with admiration at the perfection of its apparent design. But it is not real design, because it has been arrived at by a completely different process.

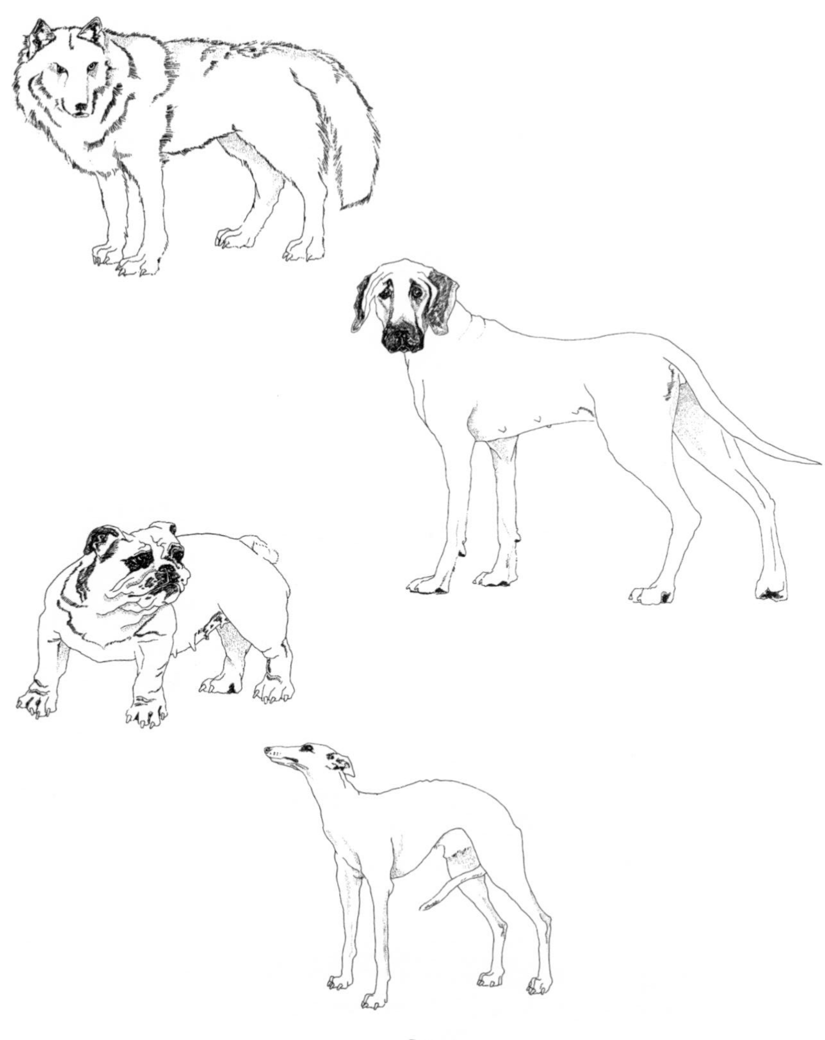

Figure 1.14 The power of artificial selection to shape animals. All these domestic dogs have been bred by humans from the same wild ancestor, a wolf (top): Great Dane, English bulldog, whippet, long-haired dachshund and long-haired chihuahua.

It would be nice to be able to demonstrate this process whenever we wish. Dog generation times are a little shorter than ours but, even so, it takes more than a human lifetime to propel dog evolution to any notable extent. Humans have bred chihuahuas in about a ten-thousandth of the time it took nature to breed wolves from their chihuahua-sized (though not chihuahua-shaped), insectivorous ancestors who lived back when the dinosaurs died out. Even so, artificial selection of real, living creatures—at least creatures bigger than bacteria—is too slow to make an impressive demonstration for impatient and short-lived humans. You can speed the process no end with a computer. Computers, whatever their faults, are blindingly fast, and they can simulate anything that can be precisely defined, which includes reproductive processes like those of animals and plants. If you simulate heredity, that most basic condition for life, and provide for occasional random mutation, it is truly startling what can evolve before your eyes in a few hundred generations of selective breeding. I pioneered this approach in my book *The Blind Watchmaker*, using a computer program of the same name. With this program you can breed, by artificial selection, creatures called computer biomorphs.

Computer biomorphs are all bred from a common ancestor that looks like this , in very much the same sense as all the breeds of dogs were bred from a wolf. Litters of progeny with random 'genetic mutations' appeared on the computer screen and a human chose which member of each litter to breed from. This needs some explanation. First, what does it mean to speak of 'progeny', of 'genes', and of 'mutations' in the case of these computer objects? All biomorphs have the same kind of 'embryology'. They are basically built as a branching tree, or a segmental series of such trees joined up to one another. Details of the tree(s), such as how many branches there are, and the lengths and angles of the various branches, are controlled by 'genes' which are just numbers in the computer. Genes in real trees, like genes in us and genes in bacteria, are coded messages written in the language of DNA. The DNA is copied from generation to generation with great, though not perfect, fidelity. Within each generation, the DNA is 'read out' and has an influence on the shape of the animal or plant. Figure 1.15 shows how, in real trees and in computer biomorph trees, changes in just a few

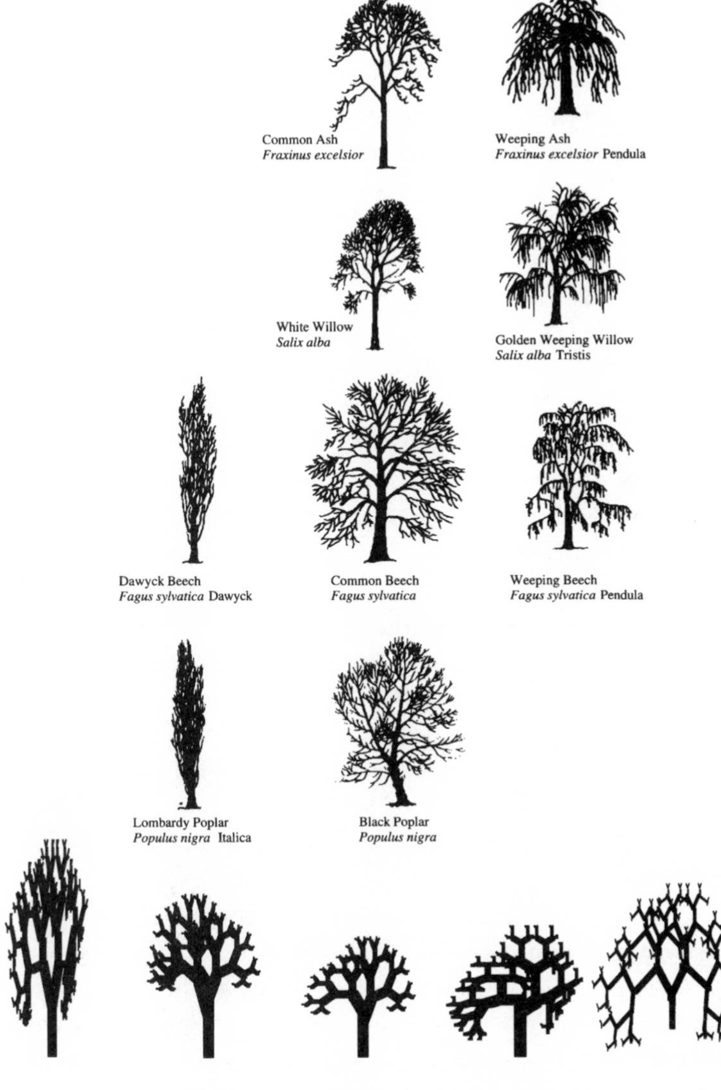

Figure 1.15 Real trees and computer biomorph trees to show how varieties of the same species can vary in shape because of quite minor changes in growth rules. Several species of tree have a weeping variety and several species have converged upon the skypointing, 'Lombardy' form.

genes can alter the shape of the whole plant by changing the programmed rules of growth as each new twig is budded off. Biomorph genes are not made of DNA but this difference is trivial for our purposes. DNA is digitally coded information just like numbers in a computer, and numerical 'genes' pass down the generations of biomorphs in the same kind of way as DNA passes down generations of plants or animals.

When a biomorph has a child, the child inherits all the genes of its parent (it only has the one parent, for there is no sex), but with some possibility of random mutation. A mutation is a slight random increase or decrease in the numerical value of a gene. So a child might be like its parent but with a slightly steeper angle to one of its branches because the numerical value of its Gene 6 has increased from 20 to 21. When in biomorph breeding mode, the computer draws a biomorph in the centre of the screen, surrounded by a litter of randomly mutated offspring. Because their genes have changed only slightly, the offspring always bear a family resemblance to the parent, and to each other, but they often display slight differences that the human eye can detect. Using the computer mouse, a human chooses one of the screenful of biomorphs for 'breeding'. The screen goes blank except for the chosen biomorph, which glides to the parental slot at the centre of the screen and then 'spawns' a new litter of mutant offspring around itself. As generations go by, the selector can guide evolution in very much the same way as humans guided the evolution of domestic dogs, but much faster. One of the things that surprised me when I first wrote the program was how quickly you could evolve away from the original tree shape. I found that I could home in on an 'insect' or a 'flower', a 'bat', a 'spider' or a 'spitfire'. Each one of the biomorphs in Figure 1.16 is the end product of hundreds of generations of breeding by artificial selection. Because the creatures breed in a computer, you can whistle through many generations of evolution in a matter of minutes. A few minutes of playing with this program on a modern, fast computer gives you a hands-on, vivid feeling for how Darwinian selection works. The biomorphs in the safari park of Figure 1.16 seem to me to resemble wasps, butterflies, spiders, scorpions, flatworms, lice, and other 'creatures' that look vaguely bio-

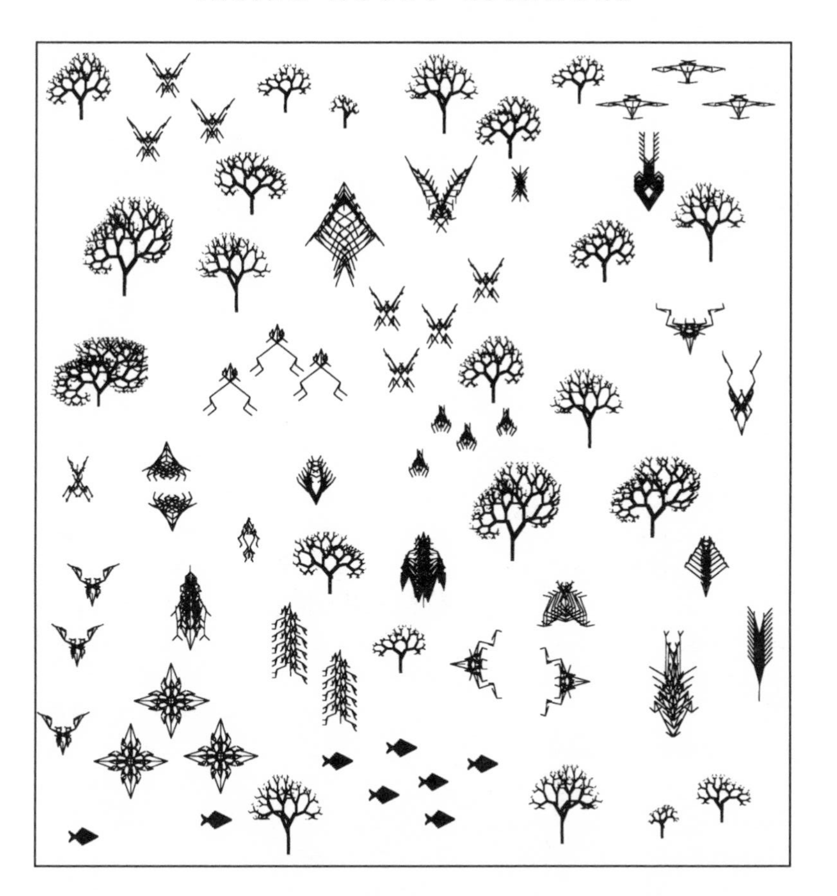

Figure 1.16 Safari park of black-and-white biomorphs, bred with the 'Blind Watchmaker' computer program.

logical even if they don't resemble *particular* species on this planet. Yet all are cousins of the trees among which they stand, and of the squadron of 'spitfires' in the top right corner. They are pretty close cousins, at that. All have the same number of genes (sixteen). They differ only in the numerically coded values of those genes. You could go from any creature in the safari park to any other, or to any of trillions of other biomorphs, simply by selective breeding.

The most recent version of the program can breed biomorphs that vary also in colour. It is based on the old program but it has a more elaborate 'embryology' and new genes controlling the colour of tree branches. There are also new genes determining whether each branch of the tree is a line or a rectangle or an oval, whether the shape concerned is filled or empty, and how thickly the lines are painted. When using the colour program I find myself following evolutionary alleys not towards insects and scorpions but towards flowers and the sort of abstract patterns that might look good on wallpaper or bathroom tiles (Figure 1.17). My wife, Lalla Ward, has embroidered four of these biomorphs as chair covers, taking precisely one needlepoint stitch for each computer pixel.

Biomorphs are 'artificially selected' by a human chooser. In this respect they are like cabbages or pedigree dogs. But artificial selection requires a human chooser and is not the main subject of this book. Following Darwin himself, I am using artificial selection as a model for a different process: natural selection. Finally, the time has come to speak of natural selection itself. Natural selection is like artificial selection, but without the human chooser. Instead of a human deciding which offspring shall die and which shall reproduce, nature 'decides'. The quotation marks are vital because nature doesn't consciously decide. This might seem too obvious to emphasize, but you'd be surprised by the number of people who think natural selection implies some kind of personal choice. They couldn't be more wrong. It just is the case that some offspring are more likely to die while others have what it takes to survive and reproduce. Therefore, as the generations go by, the average, typical creature in the population becomes ever better at the arts of surviving and reproducing. Ever better, I should specify, when measured against some absolute standard. Not necessarily more effective in practice because survival is continually menaced by other creatures who are also evolving and perfecting their arts. A species may get progressively better at the art of avoiding predators but, since predators are concurrently getting better at the art of catching prey, there may be no net gain. This kind of 'evolutionary arms race' is interesting, but we are jumping ahead of ourselves.

Artificial selection is relatively easy to achieve in the computer, and the biomorphs are a good example. It is my dream to simulate natural selection in the computer too. Ideally I'd like to set up the conditions for evolutionary arms races in which 'predators' and 'prey' emerge on the screen and goad each other into progressive evolution while we sit back and watch. Unfortunately it is very difficult, for the following reason. I said that some offspring are more likely to die, and it might seem easy enough to simulate non-random death. But, in order to be a good simulation of a natural death, the demise of the computer creature must result from some interesting imperfection, like having short legs which make it run more slowly than predators. Computer biomorphs, for instance the insect-like forms in Figure I.16, sometimes have appendages which we imagine we see as legs. But they don't use these 'legs' for anything, and they don't have predators. They don't have prey or food plants. There is no weather in their world and no disease. In theory we can simulate any of these hazards. But to model any one of them in isolation would be scarcely less artificial than artificial selection itself. We'd have to do something like arbitrarily decide that long, thin biomorphs can run away from predators better than short fat ones. It is not difficult to tell the computer to measure the dimensions of biomorphs and choose the lankiest for breeding. But the resulting evolution would not be very interesting. We'd just see biomorphs becoming more and more spindly as the generations go by. It is no more than we could have achieved by artificially selecting the spindliest by eye. It does not have the emergent qualities of natural selection, which a good simulation might achieve.

Real-life natural selection is much subtler. It is also in one sense much more complicated though in another sense it is deeply simple. One thing to say is that improvement along any one dimension, like leg length, is only improvement within limits. In real life there is such a thing, for a leg, as being too long. Long legs are more vulnerable to breaking and to getting tangled up in the undergrowth. With a little ingenuity, we could program analogues of both breakages and entanglements into the computer. We could build in some fracture physics: find a way of representing stress lines, tensile strengths, coefficients of elasticity—anything can be simulated if you know how it works. The problem comes with all the things that we don't know about or haven't thought of, and that means almost everything. Not only is the optimal leg length influenced by innumerable effects that we haven't thought of.

Worse, length is only one of countless aspects of an animal's legs that interact with each other, and with lots of other things, to influence its survival. There is leg thickness, rigidity, brittleness, weight to carry around, number of leg joints, number of legs, taperingness of legs. And we've only considered legs. All the other bits of the animal interact to influence the animal's probability of surviving.

As long as we try to add up all the contributions to an animal's survival theoretically, in a computer, the programmer is going to have to make arbitrary, human decisions. What we ideally should do is simulate a complete physics and a complete ecology, with simulated predators, simulated prey, simulated plants and simulated parasites. All these model creatures must themselves be capable of evolving. The easiest way to avoid having to make artificial decisions might be to burst out of the computer altogether and build our artificial creatures as three-dimensional robots, chasing each other around a three-dimensional real world. But then it might end up cheaper to scrap the computer altogether and look at real animals in the real world, thereby coming back to our starting point! This is less frivolous than it seems. I'll return to it in a later chapter. Meanwhile, there is a little more we can do in a computer, but not with biomorphs.

One of the main things that makes biomorphs so unamenable to natural selection is that they are built of fluorescent pixels on a two-dimensional screen. This two-dimensional world doesn't lend itself to the physics of real life in most respects. Quantities like sharpness of teeth in predators and strength of armour plating in prey; quantities like muscular strength to throw off a predator's attack or virulence of a poison do not emerge naturally in a world of two-dimensional pixels. Can we think of a real-life case of, say, predators and prey, which does lend itself, naturally and without contrived artificiality, to simulation on a two-dimensional screen? Fortunately we can. I've already mentioned spider webs when talking about designoid traps. Spiders have three-dimensional bodies and they live in a complex world of normal physics like most animals. But there is one particular thing about the way some spiders hunt that is peculiarly suited to simulating in two dimensions. A typical orb web is, to all intents and purposes, a two-dimensional structure. The insects that it catches move

Figure 1.17 Safari park of biomorphs bred by 'Colour Watchmaker'. The large black-and-white triangles in the background were added for purely decorative reasons.

in the third dimension, but at the moment of truth, when an insect is caught or escapes, the action is all in one two-dimension plane, the plane of the web. The spider web is as good a candidate as I can think of for an interesting simulation of natural selection on a two-dimensional computer screen. The next chapter is largely devoted to the story of spider webs, beginning with the natural history of real webs and moving on to computer models of webs and their evolution by 'natural' selection in the computer.

CHAPTER 2

SILKEN FETTERS

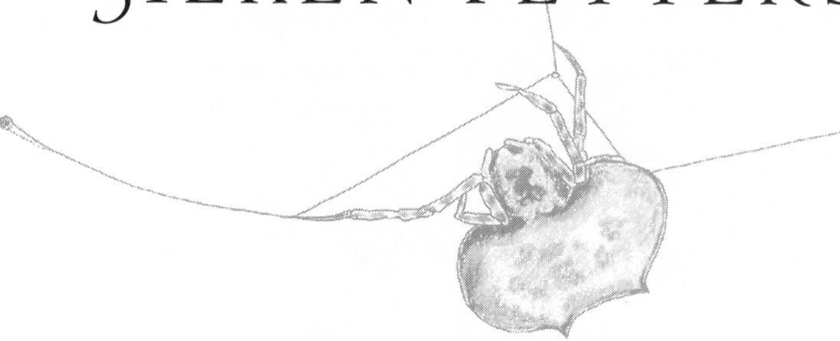

A GOOD WAY TO ORDER OUR UNDERSTANDING OF ANY living creature is to imagine, fancifully and with something more than poetic licence, that it (or, if you prefer, a hypothetical 'designer' of the creature) faces a chain of problems or tasks. First we pose the initial problem, then we think of possible solutions that might make sense. Then we look at what the creatures actually do. That often leads us to notice a new problem facing animals of this kind, and the chain continues. I did this in the second chapter of *The Blind Watchmaker,* with respect to bats and their sophisticated echo-ranging techniques. Here I shall follow the same strategy in this chapter on spider webs. Notice that the progression of problem leading to problem is not to be thought of as marching through one animal's lifetime. If it is a temporal progression at all the time scale is evolutionary, but it may sometimes be not a temporal but a logical progression.

Our fundamental task is to find an efficient method of catching insects for food. One possibility is the flying swift solution. Take to the air like the prey themselves. Fly extremely fast with the mouth open, aiming accurately with keen eyes. This method works for swifts and swallows, but it absorbs costly investment in equipment for high-speed flying and manoeuvring and a high-tech guidance system. The same is true of the bat solution, which is the nocturnal equivalent using sound echoes instead of light rays for guiding the missile.

A completely different possibility is the 'sit and wait' solution. Mantises, chameleons and certain other lizards that have evolved independently and convergently to be like chameleons make a go of this solution by being highly camouflaged and by moving in an agonizingly slow and stealthy manner until the final, explosive strike with arms or tongue. The reach of the chameleon's tongue enables it to catch a fly anywhere within a radius comparable to its own body length. The reach of the mantis's grappling arms is proportionately of the same order of magnitude. You might think that this design could be improved by lengthening the radius of capture even further. But tongues and arms that were much longer than the body's own length would be prohibitively costly to build and maintain: the extra flies they'd catch wouldn't pay for them. Can we think of a cheaper way to extend the 'reach' or radius of capture?

Why not build a net? Nets have to be made of some material and it won't be free. But unlike a chameleon's tongue the net material doesn't have to move, so doesn't need bulky muscle tissue. It can be gossamer-thin and can therefore, at low cost, be spun out to cover a much larger area. If you took the meaty protein that would otherwise have been used up in muscular arms or tongue, and reprocessed it as silk, it would go a very long way, much further than the reach of a chameleon's tongue. There is no reason why the net should not occupy an area 100 times that of the body, yet still be economically made out of secretions from small glands in the body.

Silk is a widespread commodity among arthropods (the major division of the animal kingdom to which both insects and spiders belong). Stick caterpillars belay themselves to a tree with a single thread of the stuff. Weaver ants stitch leaves together using silk extruded by their larvae, held in their jaws as living shuttles (Figure 2.1). Many caterpillars swaddle themselves in a cocoon of silk before growing into a winged adult. Tent caterpillars smother their trees with gossamer. A single domestic silkworm spins nearly a mile of silk when it builds its cocoon. But although silkworms are the basis of our own silk industry, it is really spiders that are the virtuoso silk producers of the animal kingdom, and it is surprising that spider silk is not more used by humanity. It is used for making precision cross-hairs in

Figure 2.1 Workers with silk. Weaver ants using larvae as living shuttles. *Oecophila smaragdina* from Australia.

microscopes. In his beautiful book *Self-Made Man,* the zoologist and artist Jonathan Kingdon speculates that spider silk may have inspired human children to invent one of our most vital pieces of technology, string. Birds, too, recognize the good qualities of spider silk as a material: 165 species (belonging to twenty-three independent families, which suggests that it has been discovered many times independently) are known to incorporate spider silk into the fabric of their nests. A typical orb-weaving spider, the garden cross spider *Araneus diadematus* produces six different kinds of silk from its rear-end nozzles, made in separate glands in its abdomen, and it switches between the different types for different purposes. Spiders used silk long before they evolved the ability to build orb webs. Even jumping spiders, who never build webs, leap into the air with a silk safety line attached, like mountaineers roped to their most recent secure foothold.

Silk thread, then, is anciently available in the spider tool-kit, and it is eminently suited to the weaving of an insect-catching net. We can think of a net as a means of being in lots of places at once. On its own scale, the spider is like a swallow with a whale's gape. Or like a chameleon with a fifty-foot tongue. A spider web is superbly economical. Whereas a chameleon's muscular tongue surely accounts for a substantial fraction of its total body weight, the weight of silk in a spider's web—all twenty metres of it in a big web—is less than a thousandth part of the weight of the spider's body. Moreover, the spider recycles silk after use by eating it, so very little is wasted. But net technology raises problems of its own.

A non-trivial problem for a spider in its web is to make sure that the prey, after hurtling into the web, sticks there. There are two dangers. The insect could easily tear the web and shoot straight through. This problem could be solved by making the silk very elastic, but this aggravates the second of the two dangers: the insect now bounces straight back out of the web as if from a trampoline. The ideal silk, the fibre of a research chemist's dreams, would stretch a very long way to absorb the impact of a fast-flying insect; yet at the same time, to avoid the trampoline effect, would be gently buffered in recoil. At least some kinds of spider silk have just these properties, thanks to the remarkably complicated structure of the silk itself, elucidated by Professor Fritz Vollrath and his colleagues at Oxford, and now at Aarhus, Denmark. The silk shown enlarged in Figures 2.2 and 2.3 is actually much longer than it looks, because most of its length is coiled up inside watery beadlets. It is like a necklace whose beads contain reeled-in surplus thread. The reeling in is done by a mechanism not fully understood, but the result is not in doubt. The web threads are capable of stretching out to ten times their resting length, and they also recoil slowly enough not to bounce the prey out of the web.

The next feature that the silk needs, in order to keep the prey from escaping, is stickiness. The substance that coats the silk in the reeling-in system we've just been talking about is not just watery. It is also sticky. One touch, and it is hard for an insect to escape. But not all spiders achieve stickiness in the same way. A different group called the cribellate spiders produce multi-stranded silk from a special silk gun called a cribellum. The spider then combs out the multi-stranded

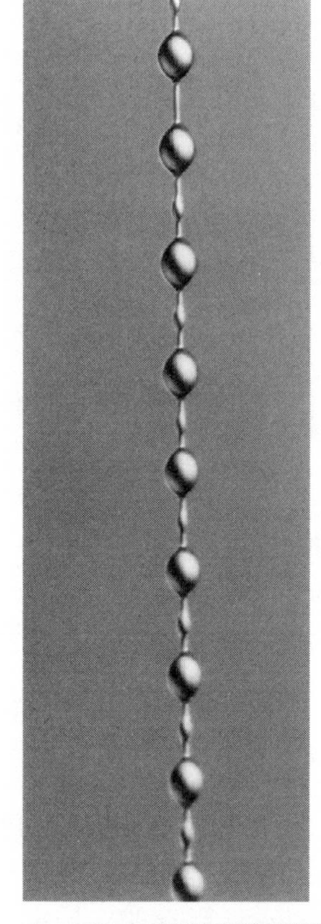

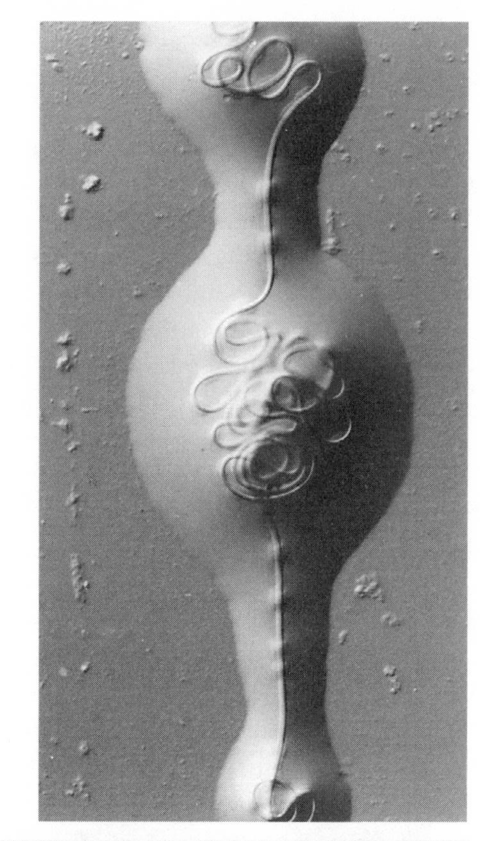

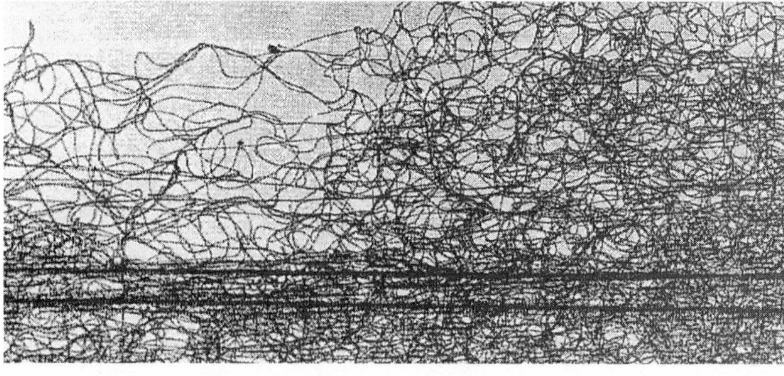

Figure 2.2 Beadlets along silk thread of spider web.
Figure 2.3 One beadlet enlarged to show coiled-up thread inside, acting as a 'windlass'.
Figure 2.4 An alternative way for a web to be sticky: hackled thread from a cribellate spider.

silk by passing it through a custom-built comb mounted on the spider's shin. Multi-stranded silk that is 'hackled' in this way puffs out into a tangly thicket (Figure 2.4). The entanglement is too small to see with the naked eye but it is just right for snagging insect legs. Hackled 'cribellate' threads behave as if they were sticky, like the gluey threads that we dealt with before. They just achieve their stickiness in a different way. In one respect, cribellate spiders have an advantage. Their threads remain sticky for longer. The non-hackling, glue-using spiders have to rebuild their gluey web anew every morning. Admittedly—and almost incredibly—this can constitute less than an hour's work, but every minute counts when you face natural selection.

But now, sticky threads pose a new and an ironic problem. Whether coated with glue or hackled into a tangle, threads sticky enough to snare an insect are tricky for a spider herself to negotiate. Spiders have no magic immunity, but evolutionary technology has come up with a mixture of partial solutions to the 'own goal' hazard. The legs of glue- using spiders are anointed with a special oil which provides some protection from the stickiness. This has been demonstrated by dipping spiders' legs in ether, which strips off the oily shield and with it the protection. A second partial solution that spiders have adopted is to make some of the threads non-sticky, namely the main spokes that radiate out from the centre of the web. The spider herself runs about on these main spokes only, using specially modified feet ending in little claws to grip the fine threads. (Male spiders build webs too. For an explanation of my sexist language, see p. 48.) She avoids the sticky spiral that winds round and round on top of the scaffolding made by the spokes. This is easy to do, because she normally sits and waits at the hub of the web, so the shortest distance to any point on the web would be along a spoke anyway.

Let's turn now to the series of problems that face a spider in actually building her web. Not all spiders are the same and, where it matters, I shall take the familiar garden spider *Araneus diadematus** as

*I shall be using Latin names, and I hope I shall be forgiven a schoolmasterly footnote on the conventions governing them because surprising numbers of educated people (perhaps the same people as wince-makingly refer to Darwin's masterwork as *Origin of the Species*) get them

(continued)

representative. Our—the spider's—initial problem is how to lay the first thread across the gap, say between a tree and a rock, where the web is to be sited. Once the gap has been spanned by that vital first thread, the spider can use it as a bridge. But how to build the first bridge? The pedestrian way would be to walk down, round, and all the way back up again, dragging a line. Spiders sometimes do this, but isn't there a more imaginative solution to the problem? Let's fly a kite. Couldn't we somehow exploit the light and airy properties of silk itself? Yes. Here's how a spider does it if there is enough wind. She releases a single thread with, at its tip, a tiny flattened silken sail or kite. This catches the air and floats. The kite is sticky and, if it happens to land on a firm surface the other side of the gap, it adheres. If the kite does not make a touch the spider hauls it back, recycles the precious silk by eating it, and tries again with a new kite. Eventually a serviceable bridge is thrown across the gap and the spider secures her own end of the thread by sticking it down. The bridge is now ready for crossing.

This first bridge is unlikely to be taut because the length of the thread will be whatever it chances to be: it is not tailor-made for the particular gap. The spider can now either shorten it to serve as one edge of the web; or she might drag it down into a V to form two of the major spokes of the web. The problem here is that, although it could be pulled down into a V, the V is unlikely to be deep enough to make two respectably long spokes. The spider's own solution to this problem is not to change the bridge itself but to use it as a support while she replaces it by a new and longer thread. Here is how she does

---

wrong. Latin names have two parts: a generic name (e.g. *Homo* is a genus) followed by a specific name (e.g. *sapiens* is the only surviving species of *Homo*), both written in italics or underlined. Names of larger units are not italicized. The genus *Homo* belongs to the family Hominidae. Generic names are unique: there is only one genus *Homo*, only one genus *Vespa*. Species often share a name with species in other genera, but there is no confusion because of the uniqueness of the generic name: *Vespa vulgaris* is a wasp, in no danger of being mistaken for *Octopus vulgaris*. The generic name always begins with a capital letter and the specific name never does (nowadays, although the original convention was that it could if derived from a proper name. Even *Darwinii* would nowadays be written *darwinii*). If ever you see (and you often will) Homo Sapiens or homo sapiens it is always a mistake. Note, by the way, that the word 'species' is both singular and plural. The plural of genus is genera.

it. Standing at one end of the bridge, she initiates a new line from her rear end, and fastens it down securely. Then she severs the existing bridge by biting it through, keeping hold of the cut end in her feet. She walks across, supported by the remains of the cut bridge in front, and by her new line which she pays out behind. She is a living link in her own bridge, moving steadily across its span. As for that part of the original bridge that she has already crossed, it has served its purpose, so she eats it. In this astonishing fashion, eating her old bridge as she goes along and creating a new one behind, she crosses from one side to the other. Moreover, her rear end is paying out silk at a faster rate than her front end is eating it. So the new bridge is, in a carefully controlled fashion, longer than the old one. Now securely fastened at both ends, it sags down the right distance to be pulled into a V and form the hub of the web.

To do this, she moves back to the centre of the new bridge and her own weight pulls it from a sagging curve to a taut V. The two arms of the V are well placed to make two of the major spokes of the web. There is little doubt about which is the next spoke to build. Clearly it would be a good idea to drop a perpendicular down from the point of the V in order to secure the future hub in place from below and keep the V taut even when the spider's own weight is not at the point. The spider fixes a new thread to the point of the V, and reels herself down like a plumb-line to the ground, or some other suitable surface, where she fastens the vertical thread. The three major spokes of the web are now neatly in place, and it looks like a Y.

The next two tasks are to put in the rest of the spokes radiating out from the centre, and the outer frame round the edge. The spider often ingeniously manages to combine the two at the same time, using staggeringly cunning techniques of wielding double and even triple threads, which are later dragged apart as the spider walks along the existing spokes. In the original draft of this chapter I explained exactly how this cat's-cradle wizardry is performed but it made my head spin to do so. When one of my editors complained that it made his head spin to read it, I was reluctantly persuaded to leave it out. The upshot of this phase of the spider's operation is a complete wheel with twenty-five or thirty spokes (the number varies from species to species, and from individual to individual), and the basic skeleton of the web is in place. But the web is still, like a bicycle wheel, mostly empty space which a fly could pass right through. Even if the fly did hit one of the threads it would not be caught because they are not sticky. What is needed now is lots of threads passing across the radial spokes. There are various ways in which these could be inserted. For instance, the spider could deal with each gap between spokes in turn, zigzagging from side to side as she makes her way from the hub to the rim, then turning and filling the next gap, and so on. But this would involve numerous changes of direction, and changes of direction waste energy and time. A better solution is to go round and round the web in a spiral, and this is what spiders mostly do, although they also double back occasionally too.

But, whether you zigzag or spiral, there are other problems. Laying the sticky thread that is actually going to do the business of catching insects is a precision matter. The spacing of the mesh must be just right. The junctions with the radial spokes must be deftly positioned so that the spokes are not pulled into an ugly mess leaving holes for prey to fly through. If the spider tried to achieve this delicate positioning while balancing on the spokes alone, her own weight would be likely to pull them out of true, and the sticky spiral thread would be joined in the wrong place with the wrong tension. Moreover, near the outer rim of the web the gap between spokes will often be too big for the spider's legs to span. Both these problems might be reduced by starting the spiral at the hub and working outwards. Near the hub the gaps are narrow, and the spokes less liable to distortion by the spider's weight because they support each other. As you spiral outwards the gaps between spokes necessarily widen, but no matter: as you come to lay each ring of the spiral, the previous, inner ring offers bridging support between the widening gaps. But the trouble with this idea is that the type of thread that is good for catching insects is very thin and elastic. It doesn't offer much support. When the whole spiral is eventually in place the web is quite robust but, during the construction process, we are talking about an incomplete and therefore weak web.

This is the main problem with laying the fine, capture spiral, but it is not the only one. Remember that, although the radiating spokes are non-sticky and relatively friendly to spiders' feet, we have now moved on to talk about the sticky silk which is specifically designed to trap prey. We have already seen that spiders are not totally immune to the stickiness of their own webs. Even if they were, using each turn of the spiral as a bridge while building the next turn might rob it of some of its precious stickiness. So, although it seems like a neat idea to build the sticky spiral from the hub outwards, walking on the previous ring, there may be, literally and metaphorically, a catch.

The spider is equal to these difficulties. Her solution is one that might occur to a human builder: temporary scaffolding. She *does* build a spiral from the hub outwards. But it is not the final, sticky gossamer, trapping spiral. It is a special, 'auxiliary' spiral which she uses once only, to help her subsequently build the sticky spiral. The auxiliary spiral is not sticky, and it is rather more widely spaced than the eventual sticky spiral. It wouldn't do for catching insects. But it is stronger than the sticky spiral will be. It stiffens and supports the web, and it gives the spider safe conduct between spokes when she finally comes to build the authentic, sticky, spiral. The auxiliary spiral takes only seven or eight turns round the web to reach from hub to rim. Having completed it, the spider switches off her non-sticky silk glands and unmasks her serious batteries: the spigots that specialize only in deadly sticky silk. She retraces her spiral steps back from rim to hub, moving in tighter and more uniformly spaced coils than on the outward journey. She uses the temporary, auxiliary spiral not only as scaffolding and support but as a sighting (actually feeling) guide. And as she goes she cuts the auxiliary spiral stage by stage, after each stage has served its purpose. As each spoke is crossed, the new, fine, sticky spiral is carefully joined to it, often in an elegant junction reminiscent of those in chicken-wire or in a fisherman's net. The temporary scaffolding is not wasteful of silk, by the way, because the fragments of it remain attached to the spokes where they are later eaten, along with the rest of the web when the spider eventually dismantles it. She doesn't eat the auxiliary silk immediately, presumably because to detach the individual fragments from the spokes would waste time.

When the spider reaches the hub on her in-going spiral journey, the web is all but complete. There is some adjusting of tensions to be done: a skilled, precision job rather like tuning a stringed instrument. She stands at the hub and tugs gently with her legs to feel the tension, makes any little lengthenings or shortenings that seem necessary, then turns and repeats the manoeuvre from another angle. Some spiders knit a complicated piece of crochet work around the hub, which may be used to fine-tune the tension in the web.

Mention of stringed instruments prompts a masculine digression. I have referred to spiders as 'she' in this story, not because males don't build webs—they do, and even new-born spiderlings can build miniature webs—but because females are larger and more prominent. Couple the larger size of females with the fact that spiders, of any age or sex, tend to eat anything smaller than themselves that moves, and it does raise problems for males. Spiders are food for beetles, ants, centipedes, toads, lizards, shrews and many birds. Whole groups of wasps are specialized to catch nothing but spiders and feed them to their larvae. But probably the most important predators of spiders are other spiders, and they are no respecters of species boundaries. Any spider who ventures on to the web of a larger spider is in mortal danger, but this is a danger that a male must face if he is to do what he has to do.

Exactly how the male copes with the problem varies from species to species. In some cases he wraps a fly in a silken parcel and presents it to the female. He waits until her fangs are safely sunk in the fly before he goes to work on her with his sexual apparatus. Males without a fly parcel may be eaten. On the other hand, males sometimes get away with presenting an empty parcel, or snatching the food out of her jaws and absconding with it after mating, perhaps to offer it to another female. In other species, the male relies on the fact that, immediately after a spider has moulted and before her new shell has hardened, she is more or less defenceless. This is when, if ever, there is a tide in the affairs of male spiders, and in several species copulation takes place at no other time than immediately after the female has moulted and she is soft and compliant, or at least disarmed.

Other species use a more appealing technique, the one that prompted my digression. Web spiders inhabit a world of silken tension. Silk lines are like extra limbs, questing antennae, almost like eyes and ears. Events are perceived via a language of tightenings and loosenings, stretchings and relaxations, shifting balances of tension. The female's heart-strings are of taut, well-tempered silk. If a male wishes to woo her and avoid, or at least postpone, being eaten, he had best play upon those strings. Orpheus himself had not better cause. In some cases the male stations himself right at the edge of the female's web and plucks the web as one might pluck a harp (Figure 2.5). This rhythmic twanging is something that no insect prey ever does, and it seems to propitiate the female. In many species the male increases his distance from the female's web by attaching to it a special 'mating thread' of his own. He plucks this special string, like a jazzman with a one-string tea-chest bass. The vibrations are transmitted along the mating thread and resonate around the female's web. They suppress, or delay, her normal urge to feed, and they lure her to walk out along the mating thread to the source of the twanging, where mating takes place. The end of the story is not always happy for the male's mortal body, but his immortal genes are by now safely stowed away inside the female. The world is well supplied with spiders whose male ancestors died after mating. The world is bereft of spiders whose would-be ancestors never mated in the first place.

Before leaving this discussion of sex and silk, make what you will of the following story. There are species of spider in which the male ties

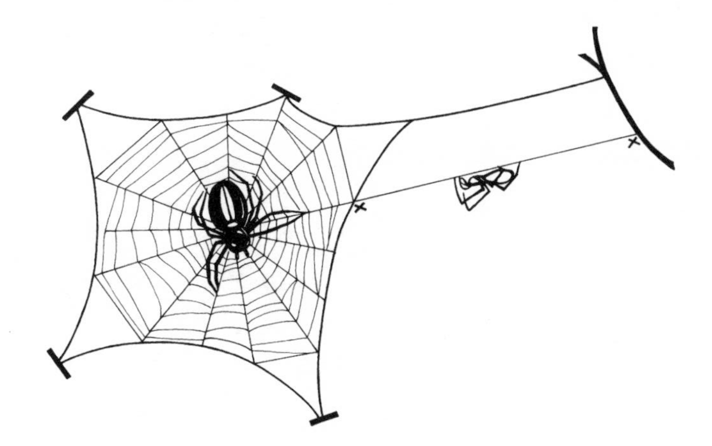

Figure 2.5 Discretion: male spider with mating thread attached to female's web.

the female down, Gulliver-like, with silk before mating with her (Figure 2.6). One is tempted to guess that he is taking advantage of the temporary eclipsing of the female's prey-catching instinct by her sexual drive, roping her down so that he can make good his escape when her feeding urge returns. I tell the tale that I heard told: the fact is that, after mating, the female has no difficulty in shaking off her fetters and striding off alone. Perhaps the ritual bondage is a symbolic vestige of an ancestral roping in earnest. Or maybe the female is hampered for just long enough to give the fleeing male a head start. He cannot, after all, want her to be tied down forever: she has to be free to lay her eggs, otherwise the whole dangerous enterprise was genetically in vain.

Let's return to the main topic of orb webs and how they are built and used. We left the web-building spider in the centre of her web at the end of the construction process, giving her attention to the fine-tuning. To carry on with our list of problems and solutions, a mesh that is fine enough to catch insects is too fine to allow the spider herself to pass from one side of the web to the other. The long detour to the edge of the web is often avoided by the simple device of a 'free zone'. This is usually a ring around the hub left free of sticky spiral thread. In some species, for instance of the genus *Zygiella*, the web has a single radial segment left vacant. Although I have introduced this hole as if it were a conduit from one side of the web to the other, it may be less important for this purpose than you might think, for *Zygiella* does not sit at the hub as many spiders do. She sits in her own

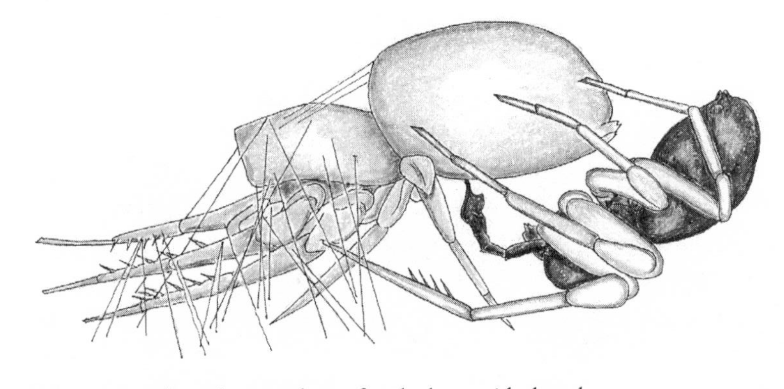

Figure 2.6 Male spider tying larger female down with threads.

tubular retreat off to one side, for a reason which brings me to my next spider problem.

Spiders, as we have seen, are not invulnerable to being eaten themselves, for instance by birds. Except in certain angled lights, or when dew-spangled, a web itself is quite hard to see because it is so fine. Its constructor, sitting four-square in the middle, is usually its most conspicuous feature. When you are fat and conspicuous to birds, there is much to be said for sitting off your web. On the other hand, it is in the nature of a spider's way of hunting that she sits and awaits prey for long periods, and the hub is the obvious place to sit because it is the junction of all the arterial trunk roads of non-sticky silk. Such a dilemma invites compromise, and different species compromise severally. Our *Zygiella* female may sit off the web, but she is never far from the centre of things. She keeps in touch by a special signal thread running from her retreat to the hub. The signal thread is under tension and it instantly transmits vibrations to the waiting spider. She races along the signal thread to the hub at a moment's notice, and from there up whichever arterial spoke will take her closest to the struggling target. The signal thread lies right down the middle of the open segment that I have already mentioned. To re-open the question of why it is open, perhaps the presence of a ladder of sticky threads would hamper her in her lightning dash to the centre of the web. Perhaps the signal thread would transmit vibrating messages less efficiently if hamstrung with crosswires.

To sit completely off the web is the compromise chosen by *Zygiella*, who doubtless pays the price of being a little slower off the mark when prey is struggling in the web (if you wonder why speed is important, we'll come to that soon). Another compromise is to sit it out at the hub, but to try to look as inconspicuous as possible. Spiders often build a dense mat of silk at the hub, behind which they can hide, or against which they can appear camouflaged. Some webs have a stripe or stripes of extra-densely zigzagging silk which might divert attention from the spider herself, lurking in the middle of it (but it has alternatively been suggested that such stripes actually are part of the spider's apparatus for fine-tuning the tensions in the web). Some spiders build extra silken ornaments into the web which look a bit like 'false spiders', and it has been suggested that they serve to deflect the pecks of birds. It's also been suggested, however, that they work in a very different way. They reflect ultra-violet light (invisible to us) in such a way that, to insect eyes, they might look like patches of blue sky or, in other words, holes.

I've made mention of a spider's need to race to the scene as soon as an insect is caught in the web. Why bother? Why not just wait until the insect ceases to struggle? The answer is that struggling by insects is often effective. They do sometimes manage to break free, especially big, strong insects like wasps. And even if they don't break free they can damage the web while trying. How to prevent an insect struggling, once initially caught, is the next of our spider problems.

The basic solution is brutally simple. Rush to where the insect is and bite it to death, guided by the vibrations from its struggles. If the insect stops struggling for a moment while you are searching for it, attempt to locate it by plucking radial threads and feeling, from the tensions of the different threads, which one is loaded down by an insect. Once the prey is reached, grapple with it and try to administer a lethal or paralysing injection of nerve poison. Most spiders have sharp, hollow fangs with poison glands (a few, such as the famous black widow, are dangerous to us but the majority of common spiders can't penetrate our skin and, even if they could, they don't have enough poison to harm a big animal). Once she has sunk her poison fangs in the prey, the spider usually hangs on for up to several minutes, waiting for the struggles to cease.

I described venomous biting as the basic method of subduing a struggling victim but it is not the only method. Most of the others—as we have come to expect of spiders—involve silk. Even before delivering the bite, most web spiders wrap some extra silk around the victim to supplement the web silk that will already be entangling its limbs and body. If the prey is dangerous like a wasp the spider usually smothers it in silk, swaddling it round and round, then finally pierces the white shroud with its fangs to give the poisoned *coup de grâce*.

Butterflies and moths, with their huge, scaly wings, present a special problem. The scales readily slough off. If we handle a moth our fingers become dusted with a fine powder made of scales. Shedding scales helps moths escape from spider webs, for powdering seems to neutral-

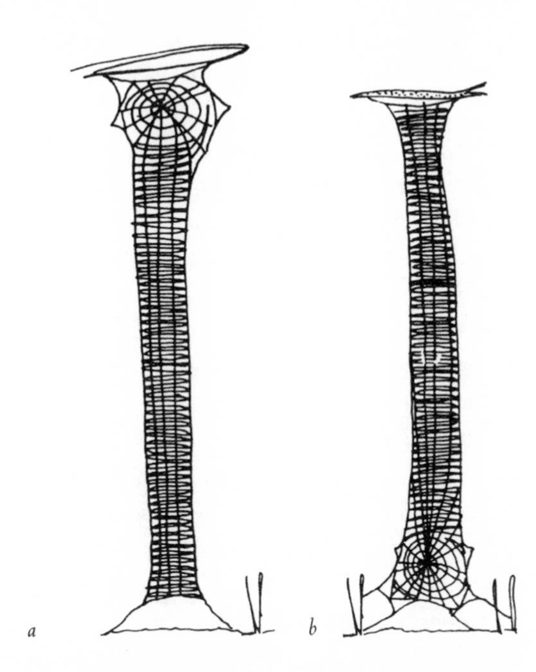

Figure 2.7 Ladder webs, independently evolved: (*a*) from New Guinea, (*b*) from Colombia.

ize the stickiness of the threads. Moths when in danger typically fold their wings and drop to the ground. Whether for this reason or simply because their wings are still partially noosed so that they can't fly, when a moth escapes from a web it will often do so by falling. This opens up a new avenue of opportunity for spiders, one that has been seized.

Michael Robinson, now Director of the National Zoo in Washington, and his wife Barbara, discovered a remarkable web in the jungle of New Guinea (Figure 2.7a). The New Guinea ladder web is basically an ordinary orb web, but with the lower side of the web stretched into a yard-long, vertical strip. The spider sits at the hub near the top. When a moth hits an orb web it has a good chance of falling free. But the New Guinea ladder spider has provided a long depth of web into which the moth can tumble. Further fluttering all the way down the web helps to use up the scaly powder and increases the chance that the moth will be detained for long enough to allow the spider to race down the ladder and thrust home its lethal bite. Soon after the Robinsons' discovery in New Guinea, their colleague William Eberhard found a New World equivalent (Figure 2.7b) in Colombia. That this ladder was invented independently of the New Guinea one is attested by a difference of detail: its hub is at the bottom of the ladder rather than the top. But it works in just the same way and apparently for the same reason: both species specialize in eating moths.

The ladder web, then, is one solution to the problem of how to impede escaping prey, one that is especially effective against moths. Another technique used by some species of spider is the sprung trap. *Hyptiotes*'s web is not a full orb, but is reduced to a triangle with only four spokes. There is an additional line attached at the point of the triangle which keeps the whole web taut. But this main guy rope, instead of being attached directly to a firm surface, is held by the spider. Indeed, the spider herself forms a living link in the attachment to a firm surface. She pulls the line taut with her front legs, and uses her third pair of legs to hold a loop of slack. She sits dangerously still in this suspended position, and waits. When an insect blunders into the web the spider reacts instantly. She releases the trap, which collapses over the insect and at the same time jerks her on towards it. She may spring the trap in two or three further stages, gathering up the slack and releasing more silk behind. The insect is now hopelessly entangled in the collapsed net. The spider wraps the victim in yet more silk and carries it away as a thickly swathed parcel. Only then does she finally bite the poor creature, inject digestive juices and slowly suck its liquefied remains out through the wall of the silken parcel. The triangular web is now not in a fit state to be used again, and it has to be rebuilt from scratch.

Presumably *Hyptiotes* is solving the problem that a web under tension, although good for catching an insect in the first place, is vulnerable to powerful struggling. If you are an insect caught by sticky threads, it is easier to pull yourself free if those threads are under tension than if they are slack. If the threads are slack, there is nothing for you to pull against and you become ever more comprehensively mired in sticky silk. Like a supersonic plane whose optimal wing shape for taking off is different from the optimum for fast flight, a spider web's optimal tension for initially catching prey is different from the optimum tension for keeping them entangled. Some planes solve their dual optimum problem by compromise: being not too bad at either task. Others—swing-wing fighter planes—get the best of both worlds by varying the geometry of their wings, albeit at a cost in complicated mechanism. *Hyptiotes* builds a variable-tension web.

Ordinary orb web spiders seem to go for the high tension that is best for initial capture, and rely on their own speed about the web to grapple the prey into submission before it can escape. Other spiders seem to go for the opposite solution, and build webs with threads that are loose in the first place (Figure 2.8). *Pasilobus* builds a triangular web with a single thread bisecting the main angle. The sticky capture threads are reduced to a few loosely hanging loops. The clever thing about these loose loops—this is another elegant discovery by Michael and Barbara Robinson in New Guinea—is that they come away specially easily at one end. An insect such as a moth, having blundered into a thread and stuck to it, quickly breaks the thread at the special low-shear junction, but remains tethered at the other end of the thread. The victim now flies round and round like a toy plane on a string. It is an easy task for the spider to haul in the line and dispatch the prey. The advantage of this arrangement may again be

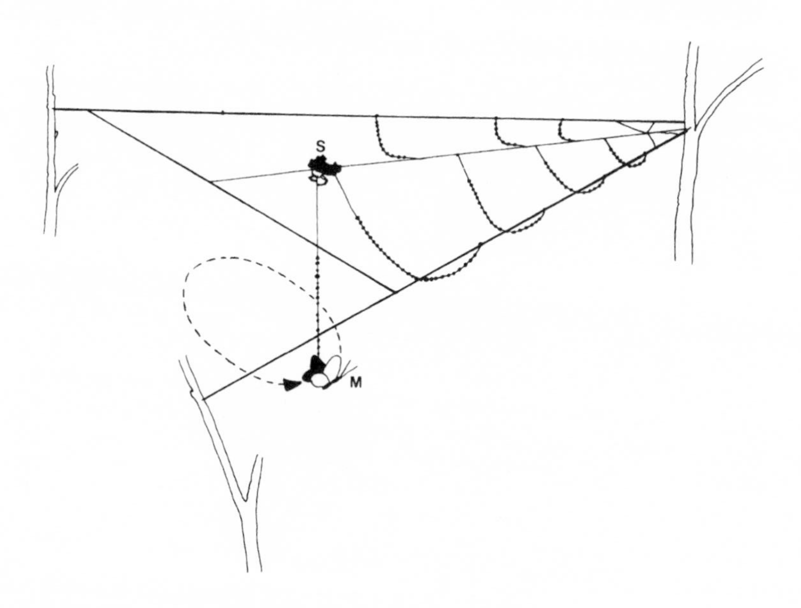

Figure 2.8 *Pasilobus* triangular web with quick-shear threads.

partly that the insect can't struggle free because everything is so loose that it can't get a solid purchase. Or it may be that the main benefit of the quick-release threads is one that harks back to an earlier member of our list of problems: how to absorb the impact of a fast-flying insect without bouncing it back like a trampoline. As in the other triangular web, it seems probable that *Pasilobus*'s triangle is the reduced descendant of a full orb web. At any rate, there is another genus, *Poecilopachys*, which uses the same quick-shear principle in a full orb web. In this case, unlike most orbs, the web is horizontal, not vertical.

If we think of *Pasilobus*'s triangle as a reduced version of *Poecilopachys*'s orb, the ultimate reduction in the same direction is the single thread of the bolas spider *Mastophora* (Figure 2.9). A bolas (or bola) is a weapon originally invented by native South Americans and still used by gauchos for hunting (for example) rheas, the large flightless birds of the pampas. It consists of a weight, such as a pair of balls or stones, on the end of a rope. It is slung towards the prey with the purpose of entangling its legs and bringing it down. The young Charles Darwin experimented with the bolas while on horseback and managed to catch his own horse—to the amusement of the gauchos though not, presumably, of the horse. The bolas spider's prey are always male moths of the family Noctuidae, and for a reason. Noctuid female moths lure their mates from a distance by releasing a unique

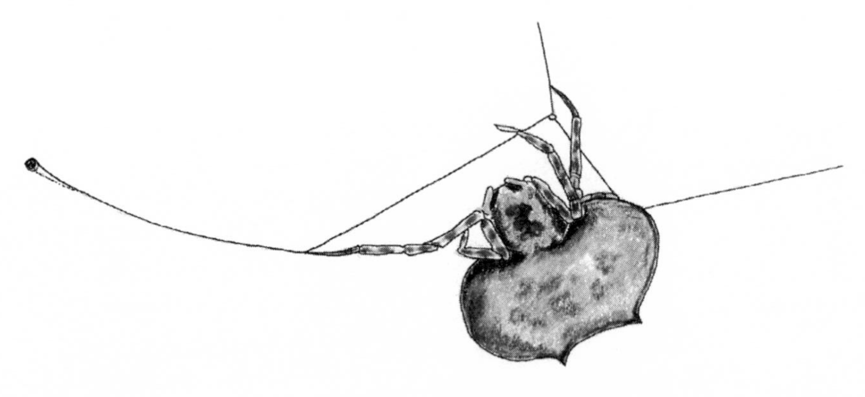

Figure 2.9 Bolas spider.

perfume. The bolas spider lures males to their deaths by synthesizing a closely similar perfume. The 'bolas' is a weighty bob on the end of a single thread of silk which the spider holds in one 'hand'. It waves the bolas around until it entangles a moth, then hauls it in. It is altogether a more high-tech affair than the gaucho's simple bag of stones. It is in fact a tightly coiled rope of silk embedded in a drop of water, like one of the sticky beadlets on an orb web. When the spider slings its bolas, the silk unreels automatically just as an angler's line unreels when he casts. If the moth is hit, it sticks, and flies round and round. The rest of the story is much the same as for the spiders with the easy-shear threads. The moth is reeled in and fanged. The bolas spider lives in South America and it is a wonderful thought that the Indians may have got the idea for their bolas by observing it in action.

We've been looking at variants on and reductions of the standard orb web. It is time to return to the orb web itself. At the end of the previous chapter we raised the question of how to take a computer model of artificial selection like the biomorph program and turn it into a model of natural selection, with blind nature instead of a human eye doing the choosing. We agreed that the snag with biomorphs was that they had nothing corresponding to a real, physical world in which to survive and be successful or unsuccessful. We could imagine some biomorphs behaving like predators; perhaps imagine them chasing other biomorphs behaving like prey. But there seems to be no natural, uncontrived way to decide which features of biomorphs will make them good, or not so good, at catching prey or at escaping predators. The human eye may see a pair of slavering, rapacious fangs mounted at one end of a biomorph (Figure 1.16, p. 33). But these gaping jaws, however fearsome they seem to our imaginations, cannot prove themselves in practice because they don't move, don't inhabit a world of real physics in which their sharpness can penetrate real shell or hide. The fangs and skin are only patterns of pixels on a two-dimensional fluorescent screen. Sharpness and toughness, brittleness and venomousness, these quantities have no meaning on the computer screen beyond contrived meanings defined as arbitrary numbers by the programmer. You can lash up a computer game in which numbers battle against other numbers, but the graphic clothing of the numbers is cosmetic and superfluous. 'Arbitrary' and 'contrived' strike the player as understatements. It was at this point at the end of the previous chapter that, with relief, we fell back on the spider web. Here was a piece of nature that could be simulated non-arbitrarily.

Orb webs in real life do their business largely in two dimensions. If the mesh is too coarse, flies pass straight through. If the mesh is too fine, rival spiders will achieve nearly the same result at less cost in silk, and will therefore leave behind more progeny to carry on their economically more prudent genes. Natural selection finds the efficient compromise. A web drawn on a computer screen has properties that interact, in ways that are hardly arbitrary at all, with flies drawn on the same screen. Size of mesh is a quantity that really means something on the computer screen, in relation to size of computer 'fly'. Total quantity of line ('cost of silk') is another such quantity. The ratio between the two that defines efficiency can be measured with an acceptably small allowance of contrived artificiality. It is even possible to import some more sophisticated physics into the computer model, and Fritz Vollrath (from whom I learned much of what I have written in this chapter), with his physicist colleagues Lorraine Lin and Donald Edmonds, have made a good start. It is easier to simulate 'elasticity' and 'breaking strain' in computer 'silk' than it is to simulate, say, 'nimbleness' in 'dodging' a computer 'predator', or 'alertness' in 'spotting' one. But in this chapter we shall be more concerned with models of web-building behaviour itself.

In writing the simulation rules for a computer spider, the programmer has the benefit of lots of detailed research on the rules actually followed by real spiders, and the decision-points that punctuate the stream of spider behaviour. Professor Vollrath and the members of his international spider research group are in the forefront of this research, and they are therefore well placed to embody the knowledge in a computer program. In fact writing a computer program is a pretty good way to summarize knowledge about any set of rules. Sam Zschokke is the member of the group who has taken on the task of summarizing, in computer form, the descriptive information about the observed movements of web-building spiders. His program is called 'MoveWatch'. Peter Fuchs and Thiemo Krink, building on work by Nick Gotts and Alun ap Rhisiart, have concentrated on the inverse task of programming 'computer spiders' catching 'computer flies'. Their program is called NetSpinner.

Figure 2.10 is MoveWatch's picture of the movements of an individual *Araneus diadematus* as she built one particular web. Note that these are not pictures of webs, although they look superficially like it. What we have here is a telescoping in time of the movements in time of a spider. It was made by videotaping the spider as she built her

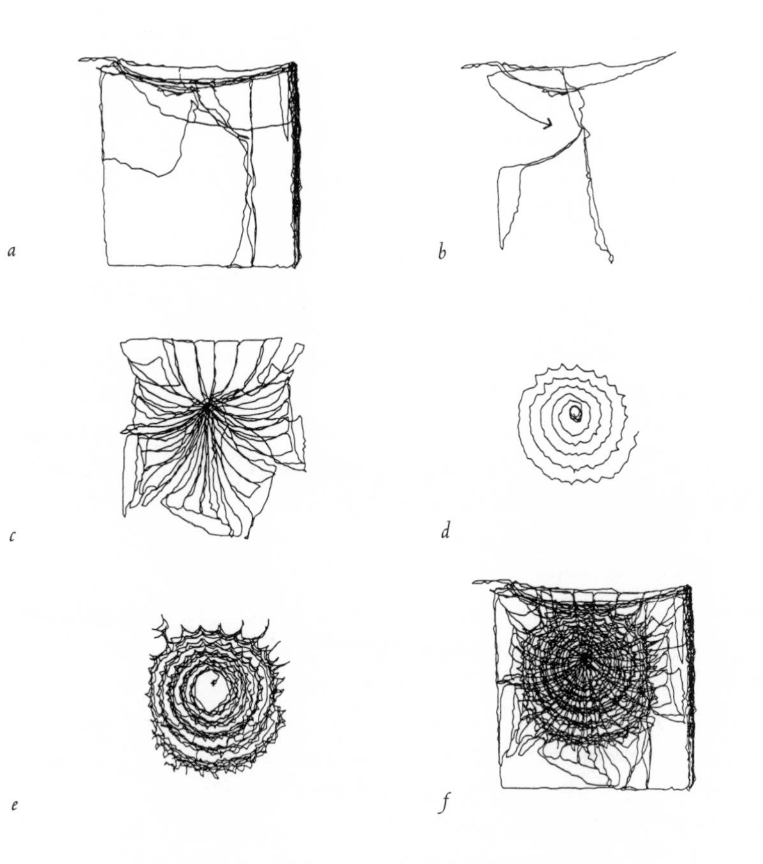

Figure 2.10 Computer-tracing of a particular (*Araneus diadematus*) spider's positions as it spins a web. MoveWatch program written by Sam Zschokke: *(a)*, *(b)* preliminaries; *(c)* radii; *(d)* auxiliary spiral; *(e)* sticky spiral; *(f)* all movements superimposed.

web. Her position at successive instants was fed into the computer in the form of a pair of grid coordinates. Then the computer drew lines between the successive positions. The 'sticky spiral' lines (Figure 2.10e), for instance, represent the trajectory of the spider while she was building the sticky spiral. They do not represent the exact positions of any silk threads. If they did, they'd be more evenly spread. As it is, they are concentrated in 'waves', reflecting the fact that the spider used the temporary, auxiliary spiral as a support while she built the sticky spiral (Figure 2.10d).

These diagrams do not represent models of the behaviour of computer spiders. Instead, they are computer descriptions of real spider behaviour. We now turn to NetSpinner, the complementary program which behaves like a kind of idealized, theoretical spider. It can be made to behave like any of a great variety of theoretical spiders. NetSpinner simulates artificial spider behaviour, in the same kind of way as the biomorph program simulated the anatomy of insect-like creatures. It builds 'webs' on the computer screen, using behavioural rules whose details vary under the influence of 'genes'. As in biomorphs, the genes are just numbers in the computer's memory, and they are transmitted forwards from generation to generation. Within each generation, the genes influence the 'behaviour' of the artificial spider and hence the shape of the 'web'. For instance, one gene might control the angle between radial spokes: mutation in this gene would change the number of spokes, by making a numerical adjustment to a behavioural rule in the computer spider. As in the biomorph program, the genes are allowed to alter their values slightly, at random, as the generations go by. These mutations show themselves as changes in web shape, and are hence subject to selection.

You can think of the six webs of Figure 2.11 as though they were biomorphs (ignore the spots for the moment). The web at top left is the parent. The other five are mutant offspring. Of course in real life webs don't give birth to webs; spiders (who build webs) give birth to spiders (who build webs). But actually there is an important sense in which what I have just said about webs here could also be said of bodies. Genes (which build human parents) give rise to genes (which build human children). In the computer model, the genes that built

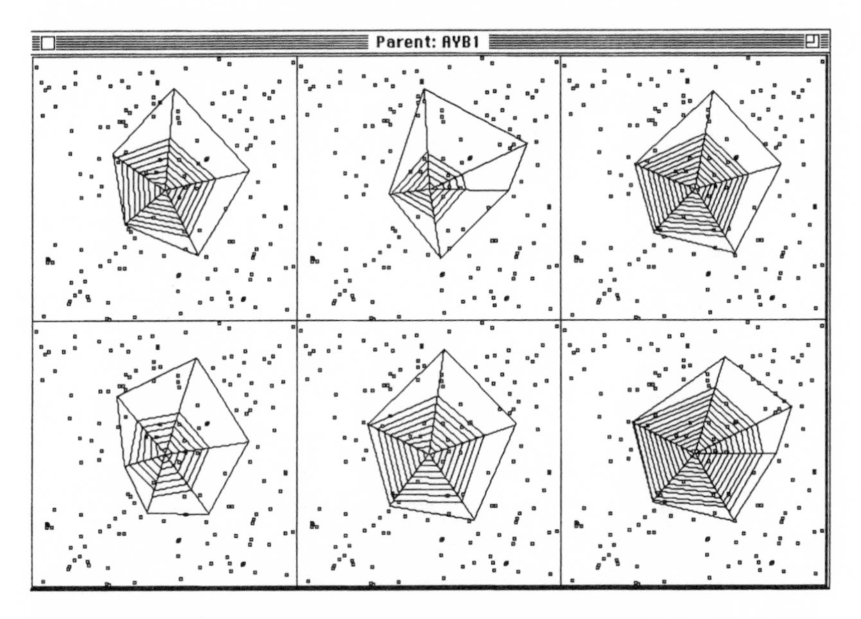

Figure 2.11 Computer-generated webs, bombarded with computer-flies. NetSpinner program written by Peter Fuchs and Thiemo Krink.

the parent web at top left (via their influence on the behaviour of a notional spider which we don't actually see on the screen) are the genes that were mutated to give rise to the genes that built the daughter webs in the other five slots.

Of course we could, as if we were choosing a biomorph for breeding, choose by eye one of the six webs for breeding. What this would mean is that *its* genes would be the ones chosen to go forward (subject to mutation) to the next generation. But that would be artificial selection. The whole point of switching from biomorphs to spider webs was that we saw a possible opportunity to simulate natural selection: selection by measured efficiency at catching 'flies' rather than selection by human aesthetic whim.

Now look at the spots on the picture. These are 'flies' that the computer has shot at random at the webs. If you look carefully, you'll notice that it's the same set of randomly positioned flies that has been shot at all six webs. This is the sort of thing that a computer, as opposed to real life, does all the time unless you go out of your way to tell it not to. It is not important in this case, and it even eases comparison between the webs. Comparison means partly that the computer counts up the number of flies 'caught' by each of the six webs. If that were all, the web at bottom right would win the contest, for its sticky spiral embraces the largest number of flies. But sheer number of flies is not the only important variable. There is also the cost of the silk. The web at top middle uses the least amount of silk, so if that were the sole criterion it would win the competition. The true winner is the web that catches the most flies *minus* a cost function computed from the length of silk. By this more sophisticated calculation, the winner is the web at bottom middle. This is the one chosen to breed and pass the genes that built it on to the next generation. As in the biomorph program, this process of breeding from winners over many generations fosters a gradual evolutionary trend. But whereas with the biomorphs the direction of the trend was guided purely by human whim, in the case of NetSpinner the direction of evolution is automatically guided towards improved efficiency. It is what we hoped for: a computer model of natural selection rather than artificial selection. And what evolves under these conditions? It is really rather gratifying how lifelike are the webs that emerge in an overnight run of forty generations (Figure 2.12).

The pictures I've shown so far were produced by NetSpinner II, which is mainly the work of Peter Fuchs (NetSpinner I was a preliminary version that I shall not discuss). Later versions of the NetSpinner program, rewritten by Thiemo Krink, steal a march on biomorphs in an additional important respect. NetSpinner III incorporates sexual reproduction. Biomorphs, and NetSpinner II, reproduced only asexually. What can it mean to say that computer spiders reproduce sexually? You don't literally see spiders copulating on the screen, though no doubt that could be managed, complete with the occasional cannibalistic climax. What the program does is arrange the genetic liaisons of sexual reproduction, the recombining of half of one parent's genes with half of the other parent's genes.

Here's how it works. In any one generation there is a population, or 'deme', of half a dozen spiders, each of whom builds a web. The shape

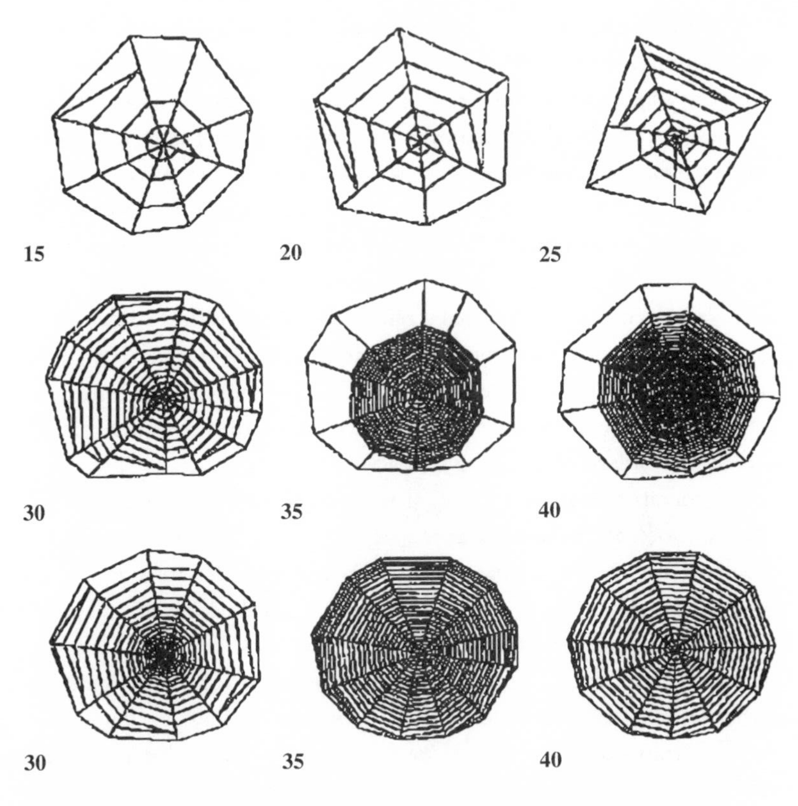

Figure 2.12 Overnight evolution of web by NetSpinner depicted every five generations.

of the web is governed by a 'chromosome' or string of genes. Each gene works by influencing a specific web-building 'rule', as we saw above. The webs are then bombarded with 'flies'. The 'goodness' of the web is calculated in the same way as before, as a function of the number of 'flies' caught minus a function of the 'silk' used. A fixed proportion of the population of spiders dies in every generation, and it is the ones with the least efficient webs that die. The remaining spiders mate with one another at random to produce a new generation of spiders. 'Mating' means that the chromosomes of the two spiders 'line up' and exchange a portion of their length. This sounds bizarre and contrived until you remember that it is exactly what real chromosomes, of ourselves as well as spiders, actually do in sexual reproduction.

The process continues and the population evolves, generation after generation, but with one further refinement. There is not just one deme of six spiders but (say) three semi-separate demes (Figure 2.13). Each of the three demes evolves in isolation except that, from time to time, an individual 'migrates' to another deme, carrying its genes with it. We'll return to the theory behind this in Chapter 4. For the moment we can briefly say that all three demes evolve towards improved webs: webs that are better at catching flies economically. Some demes may run up evolutionary blind alleys. Migrating spider genes can be thought of as an injection of fresh 'ideas' from another population. It is almost as though a successful sub-population sends out genes that 'suggest' to a less successful population a better way to solve the problem of building a web.

In generation one, in all three demes, there is a wide variety of web shapes, most of them not particularly efficient. As in the asexual example of Figure 2.12, what we observe, as the generations go by, is a gradual narrowing down of the variation towards a better and more efficient shape. But now sexual reproduction sees to it that 'ideas' are shared within demes, so the different members of each deme are rather similar to each other. On the other hand they are genetically cut off from the other demes, so there are noticeable differences between demes. At one point in generation eleven, the genes of two webs migrate across from Deme 3 to Deme 2, thereby 'infecting' Deme 2 with 'ideas' from Deme 3. By generation fifty—actually long before this in some cases—the webs have evolved to become good, stable, efficient fly-catching devices.

So, something like natural selection can work in a computer to produce artificial webs that are more efficient in catching flies than the original webs. This is still not quite true natural selection, but it is a good step closer to natural selection than the pure artificial selection of the biomorphs. But even NetSpinner still is not true natural selection. NetSpinner has to make a calculation to decide which webs are good enough to breed from and which are not. A decision has to be made by the programmer about how costly a given length of 'silk' is, in the same currency as the value of a 'fly'. The programmer could, at will, change the currency conversion rate. He could, say, double the 'price' of silk. This could lower the breeding success of larger or denser webs which, for the sake of catching a few extra flies, are a bit extravagant with silk. The programmer has to decide the currency conversion himself and he could choose any conversion factor, at will. This is only one of many such currency conversions that are going on under the surface. The rate at which fly 'flesh' is converted into baby spiders is also decided by the programmer. It could be different. The extent to which spiders die for other reasons, having nothing to do with how good their webs are, is also implicitly decided by the programmer. The decision is arbitrary, and a different decision might produce a different evolutionary result.

In real life, none of these decisions is arbitrary. None of them is really a decision at all, and no computational machinery is used to make them. They just happen, naturally and without fuss. Fly flesh just *is* converted to spider offspring flesh, and the currency conversion factor just *is*. If we come along afterwards and calculate it, that is our business. The conversion happens automatically, whether anybody puts it into mathematical economic terms or not. The same goes for the conversion of insect flesh into silk. NetSpinner, in effect, assumes that all flies are the same as each other. In real life there may be formidable complications of detail and these, too, emerge simply and without fuss. Quite apart from the fact that some insects are larger than others, there could be subtle qualitative differences. Suppose that, in order to make silk, a particular amino acid which is in short supply is necessary. Different kinds of insects vary in how rich they are in this particular amino acid. The true calculation of the value of an insect, then, has to take account of what kind of insect it is as well as how big it is. NetSpinner could compute this kind of effect, but it would be another arbitrary calcula-

Figure 2.13 (overleaf) Fifty generations of evolution of three sexually reproducing demes of computer webs bred by 'natural' selection in NetSpinner. In the eleventh generation, two web genotypes from Deme 3 migrated into Deme 2 where they were available for crossbreeding (indicated in this illustration by the solid arrows).

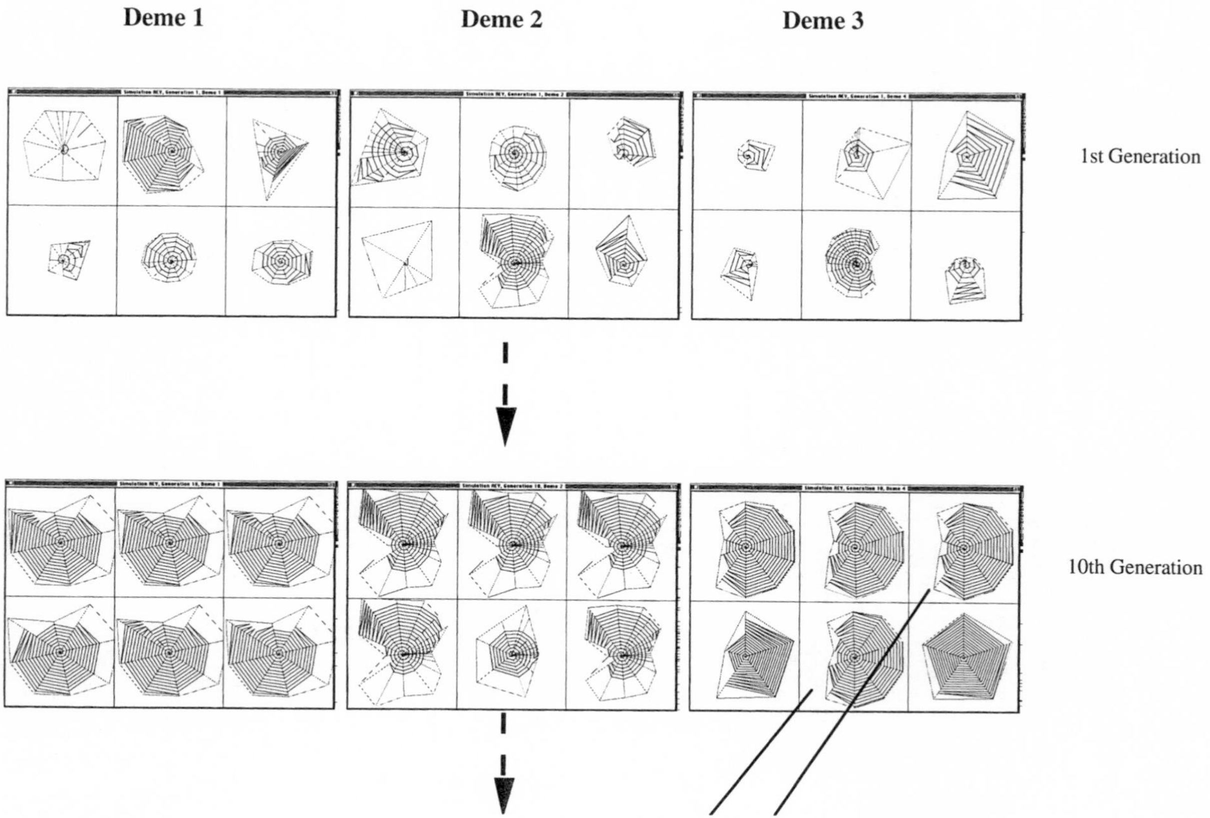

Deme 1

Deme 2

Deme 3

1st Generation

10th Generation

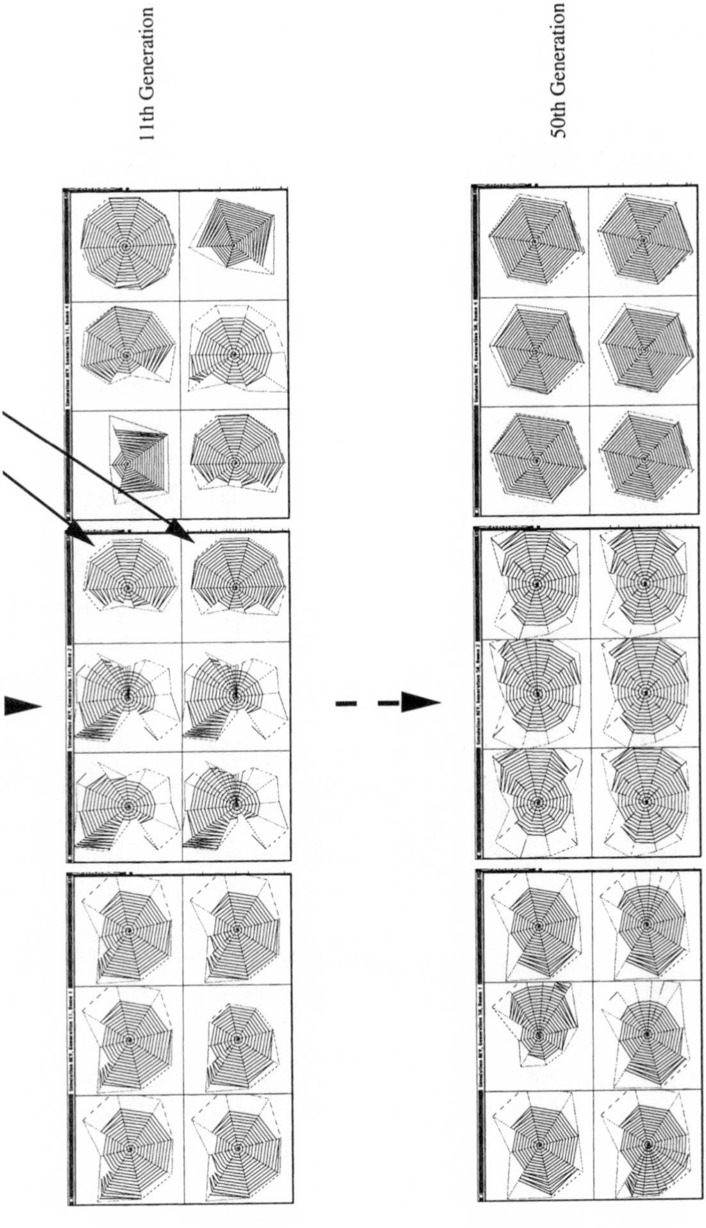

11th Generation

50th Generation

tion. In real life it just happens, automatically and without contrivance. Here's another complication. Presumably the value of one extra fly is less when a spider is nearly full than it is when she is nearly empty. NetSpinner ignores this, real life does not. NetSpinner could make an arbitrary calculation to allow for the complication of satiation. In real life it just happens, willy-nilly. No explicit calculation has to be done.

The point I am making is so obvious that it hardly needs making, yet so important that it must be made. Every time an additional and complicated point of detail is incorporated into NetSpinner, extra pages of difficult computer code have to be written by a clever human programmer. Yet in real life there is, by contrast, a marked *lack* of explicit computation. The currency-conversion factor between fly protein and silk protein is just automatically there. The fact that a fly is more valuable to an empty spider than to a full one needs no imported computation. It would be surprising if food were not more valuable to an empty spider. We are accustomed to seeing a computer model as a simplification of the real world. But there is a sense in which computer models of natural selection are not simplifications but complications of the real world.

Natural selection is an extremely simple process, in the sense that very little machinery needs to be set up in order for it to work. Of course the effects and consequences of natural selection are complex in the extreme. But in order to set natural selection going on a real planet, all that is required is the existence of inherited information. In order to set a simulated model of natural selection going in a computer, you certainly need the equivalent of inherited information, but you need a lot else besides. You need elaborate machinery for calculating lots of costs and lots of benefits and the assumed currencies for converting one to another.

More, you need to set up a whole artificial physics. We chose spider webs for our example because, of all devices in the natural world, they are among the simplest to translate into computer terms. Wings, backbones, teeth, claws, fins and feathers: in principle we could make computer models of all of them and the computer could be programmed to judge the efficiency of variant forms. But it would be an aggravatingly complicated programming task. A wing, a fin or a feather cannot show its quality unless it is placed in a physical medium—air or water—with properties such as resistance, elasticity and patterns of turbulence. These are hard to simulate. A backbone or a limb bone cannot show its quality unless placed in a physical system of stresses, leverages and frictions. Hardnesses, brittlenesses, elasticities of bending and compression—all would have to be represented in the computer. To simulate the dynamic interactions among lots of bones, strutted at various angles and roped together by ligaments and tendons, is a formidable computational task involving arbitrary decisions at every turn. To simulate air flow and turbulence around a wing is such a difficult problem that aero-engineers frequently resort to models in wind tunnels rather than attempt to simulate it on a computer.

I must not underestimate the work of computer modellers, however. The discipline of 'Artificial Life' was named in 1987 and I was honoured to be invited to the christening in Los Alamos, once the home of the atomic bomb, now turned to more constructive purposes. Christopher Langton, the inspiration and convenor of the original 1987 conference and its successors, has now founded a journal of artificial life, the first volume of which has just arrived. It contains articles that already lighten the pessimism of the previous paragraph. For example, three North Americans called Demetri Terzopoulos, Xiaoyuan Tu and Radek Grzeszczuk have written a spectacular simulation of computer fishes which behave like real fishes and interact with each other in simulated computer water. The computer world in which these fishes swim has its own simulated physics, based upon the real physics of water. Much of the programming effort goes into the simulation of a single fish, getting its behaviour right. Then this working fish is reproduced many times with variations and they are all released into the 'water' where they 'notice' each other and interact with each other. For example they avoid 'colliding' with each other and they associate with each other in 'schools'.

Each computer fish has an anatomy built up of twenty-three nodes, arranged in simulated three-dimensional space and linked to their neighbours by ninety-one 'springs' (Figure 2.14). Twelve of the springs are capable of contracting: they are the 'muscles' of the artificial fish. The sinuous swimming movements of a real fish, including turns, are simulated by controlled waves of contraction passing among the 'muscles'. The fish can learn from experience to improve the sequencing of muscular contractions to swim, turn and follow targets. Fishes have three 'mental state variables' called 'hunger', 'libido' and 'fear', which combine to generate 'intentions'. Intentions include 'eat', 'mate', 'wander about', 'leave' and 'avoid collision'. The fish has two sense organs, one that measures the 'temperature' of the water and one that acts as a crude 'eye', detecting the position, colour and size of objects out there in its world. For cosmetic purposes, the skeleton of nodes and springs is clothed in a solid-seeming, fish-coloured envelope. Different kinds of fish, for example predators and prey, are distinguished, not only by different external cosmetic renderings, but also by differences in behaviour (Figure 2.15). Predators differ from prey, not just in their size but in their behavioural predispositions, the weightings given to the three mental state variables and the various 'intentions'. Even with today's fast computers, by the way, simulations

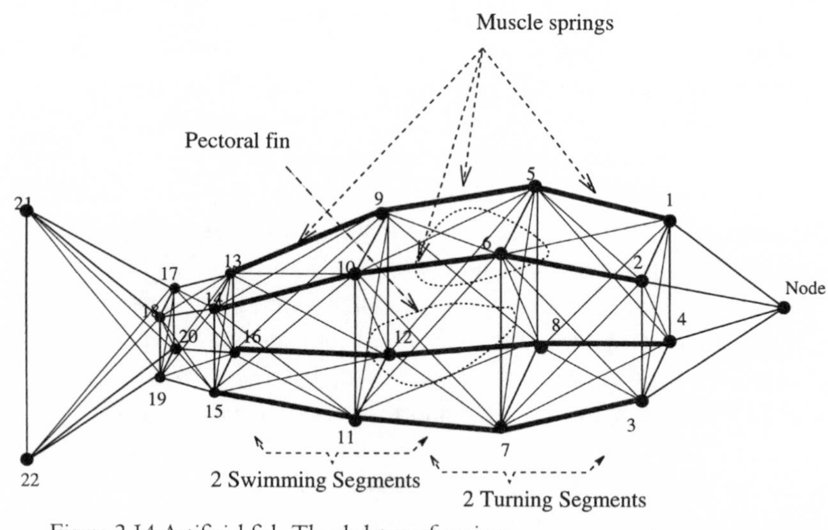

Figure 2.14 Artificial fish. The skeleton of springs.

Figure 2.15 Artificial shark stalking a school of prey fish.

of this kind are so costly in computer time that an artificial world containing many interacting fish cannot provide a plausible illusion in real time. Fish swim and chase each other, flee from each other and court each other on a slower time scale than in the real world, and we have to resort to the equivalent of time-lapse photography if we want to be entertained at life speed. This, however, is a detail, of no great theoretical importance: a problem that will disappear with future generations of computers.

Terzopoulos, Tu and Grzeszczuk's artificial world of fish in computer water is rich enough to be a good candidate for evolutionary simulation. At present, although their fish 'mate', this is limited to courtship behaviour: they do not actually reproduce. An obvious next step, of which the authors are well aware, is to set up 'genes' for the quantitative weightings of the various behavioural variables governing the muscle springs, and, at a higher level, the mental state variable and intentions. Males and females who mate could recombine their genes, with occasional mutation, to produce new generations of differing genetic constitution. Evolution by natural selection, albeit in the ultimately artifical environment of the computer, would then follow. There might be no need to define two kinds of fish called predators and prey. You might start with two species that differ only in size and mating compatibility but not in habit, and natural selection might naturally lead the larger species to evolve, over many generations, the habit of preying on the smaller. Who knows what intriguing quirks of artificial natural history might emerge before our eyes?

I foresee, and look forward to, a burgeoning field of research to which one might give the oxymoronic title Artificial Natural Selection. Nevertheless, there is a sense in which the easiest 'simulation' of the real world of natural selection is the real world itself. Bones actually do vary in breaking strain, compression elasticity, hardness, lines of force and expense in calcium consumption. You can calculate the details if you want to but, whether you calculate them or not, the fact remains that some bones break and others don't; some bones consume lots of precious calcium while others leave calcium spare to put in milk. Real life is starkly simple in this sense. Some animals are more likely to die than others. The fastest computer in America could spend a year costing out and calculating the details. But in nature the brute fact is, that some *do* die and others don't. That's all.

You can, if you wish, think of the genes in all the populations of the world as constituting a giant computer, calculating costs and benefits and currency conversions, with the shifting patterns of gene frequencies doing duty for the shuttling 1s and 0s of an electronic data processor. It is quite an illuminating insight to which we'll return in the closing pages of this book. But now it is time to illuminate the title. What is Mount Improbable and what shall we learn from it?

CHAPTER 3

THE MESSAGE FROM THE MOUNTAIN

MOUNT IMPROBABLE REARS UP FROM THE PLAIN, LOFTING its peaks dizzily to the rarefied sky. The towering, vertical cliffs of Mount Improbable can never, it seems, be climbed. Dwarfed like insects, thwarted mountaineers crawl and scrabble along the foot, gazing hopelessly at the sheer, unattainable heights. They shake their tiny, baffled heads and declare the brooding summit forever unscalable.

Our mountaineers are too ambitious. So intent are they on the perpendicular drama of the cliffs, they do not think to look round the other side of the mountain. There they would find not vertical cliffs and echoing canyons but gently inclined grassy meadows, graded steadily and easily towards the distant uplands. Occasionally the gradual ascent is punctuated by a small, rocky crag, but you can usually find a detour that is not too steep for a fit hill-walker in stout shoes and with time to spare. The sheer height of the peak doesn't matter, so long as you don't try to scale it in a single bound. Locate the mildly sloping path and, if you have unlimited time, the ascent is only as formidable as the next step. The story of Mount Improbable is, of course, a parable. We shall explore its meaning in this and the next chapters.

The following is from a letter that *The Times* of London published a few years ago. The author, whose name I have withheld to spare embarrassment, is a physicist, regarded sufficiently highly by his peers to have been elected a Fellow of the Royal Society, Britain's most distinguished learned institution.

> Sir, I am one of the physical scientists . . . who doubt Darwin's theory of evolution. My doubts arise not from any religious motive or desire to add fuel to either side of any controversy but merely because I think that Darwinism is scientifically indefensible.
>
> . . . We have no option but to accept evolution—all the fossil evidence points to it. The contention is only about the cause. Darwin maintains that the cause was chance: as generation succeeded generation there would be minor variations at random, those that gave some advantage would persist and those that did not would disappear. Thus living beings would gradually improve with, for example, enhanced powers of obtaining food or of destroying their enemies. This process Darwin called natural selection.
>
> As a physicist, I cannot accept this. It seems to me to be impossible that chance variation should have produced the remarkable machine that is the human body. Take only one example—the eye. Darwin admitted that this defeated him—he could not see how it could have evolved from a simple light-sensitive organ ... I myself can see no alternative to the hypothesis that living matter was designed. The origin of life is not explainable in terms of standard science nor is the wonderful succession of living creatures formed throughout the thousands of millions of years of this planet's existence.
>
> But who was the Designer?
>
> Yours faithfully,

The author is at pains to let us know, twice, that he is a physicist, which gives special weight to his views. Another physical scientist, a professor of chemistry at San Jose State University, California, has burst into biology with a publication called 'The Smyrna Fig requires God for its Production'. He describes the remarkable complexity of the relationship between figs and their wasp pollinators (see Chapter 10) and he comes to the following conclusion: 'A young wasp lies dormant in a caprifig all winter, but hatches at the exact time to lay her eggs in the summer crop of caprifigs which is necessary to pollinate the fruit. This all requires exact timing which means God controls it'! (The exclamation mark is mine.) 'To think that all of this exact pattern resulted from evolutionary chance is preposterous. Without God nothing like the Smyrna fig could exist . . . Evolutionists pretend that things arise by chance without a definite purpose or a completely thought out plan.'

One of Britain's most famous physical scientists, Sir Fred Hoyle (incidentally the author of *The Black Cloud*, which must be among the best science-fiction novels ever written), frequently expresses a similar view with respect to large molecules such as enzymes, whose inherent 'improbability'—that is the probability that they'd spontaneously come into existence by chance—is easier to calculate than that of eyes or figs. Enzymes work in cells rather like exceedingly numerous machine tools for molecular mass production. Their efficacy depends upon their three dimensional shape, their shape depends upon their coiling behaviour, and their coiling behaviour depends upon the sequence of amino acids which link up in a chain to make them. This exact sequence is directly controlled by genes and it really matters. Could it come about by chance?

Hoyle says no, and he is right. There is a fixed number of amino acids available, twenty. A typical enzyme is a chain of several hundred links drawn from the twenty. An elementary calculation shows that the probability that any particular sequence of, say 100, amino acids will spontaneously form is one in 20 × 20 × 20 . . . 100 times, or 1 in 20^{100}. This is an inconceivably large number, far greater than the number of fundamental particles in the entire universe. Sir Fred, bending over backwards (unnecessarily, as we shall see) to be fair to those whom he sees as his Darwinian opponents, generously shortens the odds to 1 in 10^{20}. A more modest number to be sure, but still a horrifyingly low probability. His co-author and fellow astrophysicist, Professor Chandra Wickramasinghe, has quoted him as saying that the spontaneous formation by 'chance' of a working enzyme is like a hurricane blowing through a junkyard and spontaneously having the luck to put together a Boeing 747. What Hoyle and Wickramasinghe miss is that Darwinism is *not* a theory of random chance. It is a theory of random mutation plus *non-random* cumulative natural selection. Why, I wonder, is it so hard for even sophisticated scientists to grasp this simple point?

Darwin himself had to contend with an earlier generation of physical scientists crying 'chance' as the alleged fatal flaw in his theory. William Thomson, Lord Kelvin, was perhaps the greatest physicist of his day and Darwin's most distinguished scientific opponent. Among his many achievements he calculated the age of the Earth based on rates of cooling, assuming that it had once been a part of the 'fires' of the Sun. He concluded that the Earth was some tens of millions of years old. Modern estimates put the age up in the thousands of millions of years. It is no discredit to Lord Kelvin that his estimate was one hundredth part of the right answer. Dating methods using radioactive decay were not available in his time, and nuclear fusion, the true 'fire' of the Sun, was unknown, so his cooling calculation was doomed from the start. What is less forgivable was his lofty dismissing, 'as a physicist', of Darwin's biological evidence: the earth wasn't old enough; there hadn't been enough time for the Darwinian process of evolution to have achieved the results we see around us; the evidence of biology must simply be wrong, trumped by the superior evidence of physics. Darwin might just as well have retorted (he didn't) that the biological evidence clearly indicates evolution, therefore there must have been time for evolution to occur, therefore the physicist's evidence must be wrong!

To return to the point about 'chance', Lord Kelvin used the prestigious platform of his Presidential Address to the British Association to quote, with approval, the words of another distinguished physical scientist, Sir John Herschel, who also, by the way, referred to Darwinism as 'The Law of Higgledy-Piggledy':

> We can no more accept the principle of arbitrary and casual variation and natural selection as a sufficient account, *per se,* of the past and present organic world, than we can receive the Laputan method of composing books (pushed *à l'outrance*) as a sufficient one of Shakespeare and the *Principia.*

Herschel's allusion was to *Gulliver's Travels* in which Swift had mocked the Laputan method of writing books by combining words at random. Herschel and Kelvin, Hoyle and Wickramasinghe, my anonymously quoted physical scientists and any number of Jehovah's Witness tracts all make the mistake of treating Darwinian natural selection as though it were tantamount to Laputan authorship. To this day, and in quarters where they should know better, Darwinism is widely regarded as a theory of 'chance'.

It is grindingly, creakingly, crashingly obvious that, if Darwinism were really a theory of chance, it couldn't work. You don't need to be a mathematician or physicist to calculate that an eye or a haemoglobin molecule would take from here to infinity to self-assemble by sheer higgledy-piggledy luck. Far from being a difficulty peculiar to Darwinism, the astronomic improbability of eyes and knees, enzymes and elbow joints and the other living wonders is precisely the problem that *any* theory of life must solve, and that Darwinism uniquely *does* solve. It solves it by breaking the improbability up into small, manageable parts, smearing out the luck needed, going round the back of Mount Improbable and crawling up the gentle slopes, inch by million-year inch. Only God would essay the mad task of leaping up the precipice in a single bound. And if we postulate him as our cosmic designer we are left in exactly the same position as when we started. Any Designer capable of constructing the dazzling array of living things would have to be intelligent and complicated beyond all imagining. And complicated is just another word for improbable—and therefore demanding of explanation. A theologian who ripostes that his god is sublimely simple has (not very) neatly evaded the issue, for a sufficiently simple god, whatever other virtues he might have, would be too simple to be capable of designing a universe (to say nothing of forgiving sins, answering prayers, blessing unions, transubstantiating wine, and the many other achievements variously expected of him). You cannot have it both ways. Either your god is capable of designing worlds and doing all the other godlike things, in which case he *needs* an explanation in his own right. Or he is not, in which case he cannot *provide* an explanation. God should be seen by Fred Hoyle as the ultimate Boeing 747.

The height of Mount Improbable stands for the combination of perfection and improbability that is epitomized in eyes and enzyme molecules (and gods capable of designing them). To say that an object like an eye or a protein molecule is improbable means something rather precise. The object is made of a large number of parts arranged in a very special way. The number of possible ways in which those parts could have been arranged is exceedingly large. In the case of a protein molecule we can actually calculate that large number. Isaac Asimov did it for the particular protein haemoglobin, and called it the Haemoglobin Number. It has 190 noughts. That is the number of ways of rearranging the bits of haemoglobin such that the result would not be haemoglobin. In the case of the eye we can't do the equivalent calculation without fabricating lots of assumptions, but we can intuitively see that it is going to come to another stupefyingly large number. The actual, observed arrangement of parts is improbable in the sense that it is only one arrangement among trillions of possible arrangements.

Now, there is an uninteresting sense in which, with hindsight, any particular arrangement of parts is just as improbable as any other. Even a junkyard is as improbable, with hindsight, as a 747, for its parts could have been arranged in so many other ways. The trouble is, most of those ways would also be junkyards. This is where the idea of quality comes in. The vast majority of arrangements of the parts of a Boeing junkyard would not fly. A small minority would. Of all the trillions of possible arrangements of the parts of an eye, only a tiny minority would see. The human eye forms a sharp image on a retina, corrected for spherical and chromatic aberration; automatically stops down or up with an iris diaphragm to keep the internal light intensity relatively constant in the face of large fluctuations in external light intensity; automatically changes the focal length of the lens depending upon whether the object being looked at is near or far; sorts out colour by comparing the firing rates of three different kinds of light-sensitive cell. Almost all random scramblings of the parts of an eye would fail to achieve any of these delicate and difficult tasks. There is something very special about the particular arrangement that exists. All particular arrangements are as improbable as each other. But of all particular arrangements, those that aren't useful hugely outnumber those that are. Useful devices are improbable and need a special explanation.

R. A. Fisher, the great mathematical geneticist and founder of the modern science of statistics, put the point in 1930, in his usual meticulous style (I never met him, but one can almost hear his fastidiously correct dictation to his long-suffering wife):

> An organism is regarded as adapted to a particular situation, or to the totality of situations which constitute its environment, only in so far as we can imagine an assemblage of slightly different situations, or environments, to which the animal would on the whole be less well adapted; and equally only in so far as we can imagine an assemblage of slightly different organic forms, which would be less well adapted to that environment.

Eyes, ears and hearts, the wing of a vulture, the web of a spider, these all impress us by their obvious perfection of engineering no matter where we see them: we don't need to have them presented to us in their natural surroundings to see that they are good for some purpose and that, if their parts were rearranged or altered in almost any way, they would be worse. They have 'improbable perfection' written all over them. An engineer can recognize them as the kind of thing that he would design, if called upon to solve a particular problem.

This is another way of saying that objects such as these cannot be explained as coming into existence by chance. As we have seen, to invoke chance, on its own, as an explanation, is equivalent to vaulting from the bottom to the top of Mount Improbable's steepest cliff in one bound. And what corresponds to inching up the kindly, grassy slopes on the other side of the mountain? It is the slow, cumulative, one-step-at-a-time, non-random survival of random variants that Darwin called natural selection. The metaphor of Mount Improbable dramatizes the mistake of the sceptics quoted at the beginning of this chapter. Where they went wrong was to keep their eyes fixed on the vertical precipice and its dramatic height. They assumed that the sheer cliff was the only way up to the summit on which are perched eyes and protein molecules and other supremely improbable arrangements of parts. It was Darwin's great achievement to discover the gentle gradients winding up the other side of the mountain.

But is this one of those rare cases where it is really true that there is no smoke without fire? Darwinism is widely misunderstood as a theory of pure chance. Mustn't it have done something to provoke this canard? Well, yes, there is something behind the misunderstood rumour, a feeble basis to the distortion. One stage in the Darwinian process is indeed a chance process—mutation. Mutation is the process by which fresh genetic variation is offered up for selection and it is usually described as random. But Darwinians make the fuss that they do about the 'randomness' of mutation only in order to *contrast* it to the non-randomness of selection, the other side of the process. It is not *necessary* that mutation should be random in order for natural selection to work. Selection can still do its work whether mutation is directed or not. Emphasizing that mutation *can* be random is our way of calling attention to the crucial fact that, by contrast, selection is sublimely and quintessentially *non*-random. It is ironic that this emphasis on the contrast between mutation and the non-randomness of selection has led people to think that the whole theory is a theory of chance.

Even mutations are, as a matter of fact, non-random in various senses, although these senses aren't relevant to our discussion because they don't contribute constructively to the improbable perfection of organisms. For example, mutations have well-understood physical causes, and to this extent they are non-random. The reason X-ray machine operators take a step back before pressing the trigger, or wear lead aprons, is that X-rays cause mutations. Mutations are also more likely to occur in some genes than in others. There are 'hot spots' on chromosomes where mutation rates are markedly higher than the average. This is another kind of non-randomness. Mutations can be reversed ('back mutations'). For most genes, mutation in either direction is equally probable. For some, mutation in one direction is more frequent than back mutation in the reverse direction. This gives rise to so-called 'mutation pressure'—a tendency to evolve in a particular direction regardless of selection. This is yet another sense in which mutation can be described as non-random. Notice that mutation pressure does not systematically drive in the direction of improvement. Nor do X-rays. Quite the contrary: the great majority of mutations, however caused, are random with respect to quality, and that means they are usually bad because there are more ways of getting worse than of getting better.

One could imagine a theoretical world in which mutations were biased towards improvement. Mutations in this hypothetical world would be non-random not just in the sense that mutations induced by X-rays are non-random: these hypothetical mutations would be systematically biased to keep one jump ahead of selection and anticipate the needs of the organism. But this is the one kind of non-randomness which, contrary to numerous theoretical yearnings, almost certainly has no basis in fact: mutations are not systematically likely to anticipate the needs of the organism, nor is it clear how such anticipation could possibly work. What might 'anticipation' mean? Suppose a terrible ice age is closing in on a previously temperate region and the local deer are perishing in their lightweight coats. Most individuals will die anyway but the species will be saved if only, in the nick of time, it can evolve a thick coat like a musk-ox. It is in principle possible to imagine a mechanism tuned to switch on desirable mutations as and when they are needed. We know that X-rays increase the general mutation rate, indiscriminately making thinner coats or thicker coats. What if intense cold could somehow increase the mutation rate in one direction only: towards thicker coats? And, symmetrically, what if intense heat could induce mutations in the other direction, towards thinner coats?

Darwinians wouldn't *mind* if such providential mutations were provided. It wouldn't undermine Darwinism, though it would put paid to its claims for exclusivity: a tailwind on a transatlantic flight can speed up your arrival in an agreeable way, and this doesn't undermine your belief that the primary force that got you home is the jet engine. But Darwinians would actually be pretty surprised (as well as intrigued) if any such beneficent mutational mechanism were discovered, for three reasons.

First, despite energetic searching, no such mechanism has yet been discovered (at least in animals and plants: there is a very special and not generally relevant suggestion of a case in bacteria in which the facts are still controversial). Second, there is no existing theory that could explain how the body might 'know' which sort of mutation to induce. I suppose one could imagine that, if there have been dozens of ice-age cycles stretching back over the millions of years, constituting a form of 'race experience', some as yet undiscovered kind of higher order natural selection could have built in a propensity to mutate in the right direction at the first inklings of the next ice age. But I repeat that there is no evidence for any such effect and, moreover, no theory so far worked out can handle it. Third—and this returns to my earlier point—some Darwinians, including me, find the proposed mechanism of directed mutation inelegantly superfluous. This is a largely aesthetic reaction which should therefore not be treated as overwhelming. But if we do react unsympathetically towards suggestions of directed mutation, it is because such suggestions are often made by people who mistakenly think that such a theory is *needed*: who don't understand that selection, on its own, is amply powerful enough to do the job, even if mutation is random. One way to dramatize the adequacy of non-random selection is to emphasize that mutation is *allowed* by the theory to be random. But, as I said before, it is not critical to the theory that mutation *must* be random, and it most certainly provides no excuse to tar the whole theory with the brush of randomness. Mutation may be random, but selection definitely is not.

Before we leave our deer out there in the deepening cold, there is one variant of the theory of providential mutation that might have occurred to you as you read the last three paragraphs. It may indeed be hard to see how the body could 'know' that cold weather requires mutation towards a thicker coat, while hot weather requires mutation in the other direction. But it is slightly easier to imagine that mutation rates might be pre-programmed to inflate, indiscriminately in all directions, during times when the going gets tough. The intuitive rationale would be something like this. A new crisis, like an ice age or an age of intense heat, is felt by the body as stress. High stress on me, whether from cold, heat, drought or any cause unspecified, indicates that *something* is wrong with my bodily equipment for present conditions. It may be too late for me, but perhaps life could be improved for some of my children if I mutate the genes in my sex organs wildly, in all sorts of random directions. Whatever the environmental crisis may be (cold, heat, drought, flood), those of my mutant children who contain mutations that turn out to be in the wrong direction (probably the majority) will die. But they would have died anyway if the crisis were sufficiently severe. Perhaps by producing a brood of mutant freaks and monsters an animal increases its chances of producing one child that is better at coping with the new crisis than it is itself.

There are, indeed, genes whose effect is to control the mutation rate in other genes. In theory one might argue that these 'mutator genes' could be triggered by stress, and such a tendency could be favoured by some sort of high-level natural selection. But alas, this theory turns out to have no more support than our previous theory of beneficently directed mutation. First, there is no evidence for it. More seriously, there are profound theoretical difficulties with any view of increased mutation rates that has them being positively favoured by natural selection. The argument is a general one which leads to the conclusion that mutator genes will always tend to disappear from the population, and this will apply to our hypothetical stressed animals.

Briefly, the general argument is as follows. Any animal that has succeeded in reaching an age to become a parent must already be pretty good. If you start with something pretty good and change it at random, the chances are that you'll make it worse. And, as a matter of fact, the great majority of mutations do make things worse. It is true that a small minority of mutations may make things better—that's ultimately why evolution by natural selection is possible at all. It is also true that a mutator gene, by increasing the total mutation rate, can help its possessor to come up with that precious rarity, a mutation that is an improvement. When this happens, that particular copy of the mutator gene itself will temporarily flourish, for it will share a body with the improved mutation that it has helped to create. You might think that this constitutes positive natural selection in favour of the mutator gene and therefore, by this means, mutation rates could increase. Alas, mark the sequel.

In future generations, sexual reproduction will do its shuffling work, rearranging and recombining the genes that share individual bodies. As the generations pass, there is nothing to stop the mutator gene becoming detached from the good gene that it created: some individuals will be born with only the good gene, others with only the mutator. The good gene itself will go on being rewarded by natural selection and may go from strength to strength in future populations. But the unfortunate mutator gene that created it has been cast aside by sexual reshuffling. Like any other gene, the mutator gene's long-term fate depends on its *average* effects: its effects averaged over all the different bodies in which it finds itself, over the long term. The average effects of the good gene that the mutator created are good, and the good gene will survive in more and more bodies in the population. But the average effects of the mutator itself are bad and, in spite of its occasional flashes of benefit, on average the mutator is bound to be penalized by natural selection. Most of the bodies in which it finds itself will be freaks or dead.

This argument against the possibility of mutator genes being positively selected depends upon the assumption that reproduction is sexual. If reproduction is asexual, the 'shuffling' step of the argument is missing. Mutator genes can be favoured by natural selection over long periods because, when there is no sex, they are not detached from the occasional good genes that they create, and they can 'hitch-hike' down the generations on the backs of good genes. When reproduction is asexual, a new good mutation will initiate a new clone of prospering individuals. A new bad mutation will quickly disappear, dragging its subclone of freaks down with it. If a good mutation is sufficiently good, the clone will continue to prosper, and all the genes in it will cash in—even the bad ones. The bad ones prosper because, in spite of their ill-effects, the average quality of the genes in the clone is positive. And among these prospering coat-tail-hangers will be the mutator gene responsible for creating the good mutation in the first place. As far as the good mutation is concerned, it will 'wish' it could shake off the dead weight of bad genes, and the mutator that created it is no exception. The good mutation, if it could think, would yearn for some good, cleansing sexual reproduction. If only my bodies would go in for some sex, it would say, I could shake off this crowd of ill-favoured hitch-hikers. I could be valued for my own virtues alone. Some of the bodies I found myself would be bad, others good, but *on average* I'd be free to benefit from my own good effects. The bad genes, on the other hand, have no 'desire' for sexual reproduction: they are on to a good thing. If they had to go it alone in the genetic free-for-all which is sex, they'd soon go under.

This argument doesn't, by itself, constitute an explanation for why we have sexual reproduction in the first place, although it might be made the basis for such an explanation. To say, as I have, that good genes can benefit from the existence of sex whereas bad genes can benefit from its absence, is not the same thing as explaining why sex is there at all. There are many theories of why sex exists, and none of them is knock-down convincing. One of the earliest to be suggested, 'Muller's Ratchet', is a more disciplined version of the theory that I've informally expressed in the form of 'wishes' of good genes and bad genes. My discussion of mutator genes can be seen as adding an extra fillip to the Muller's Ratchet theory. Asexual reproduction doesn't just allow bad genes to accumulate in the population. It gives active encouragement to mutator genes. This is likely to hasten the extinction of asexual clones or, in other words, speed up the operation of Muller's Ratchet. But the whole question of sex and why it is there, Muller's Ratchet and all, is another story and a difficult one to tell. Maybe one day I'll summon up the courage to tackle it in full and write a whole book about the evolution of sex.

But that was a digression. The upshot is that, where there is sexual reproduction, the phenomenon of mutation is penalized by natural selection, even though individual mutations (a minority of them) may occasionally be favoured by natural selection. This is true even in a time of stress where you can make a superficially plausible case for increased mutation rates. The predilection to mutate is always bad, even though individual mutations occasionally turn out to be good. It is best, if more than a little paradoxical, to think of natural selection as favouring a mutation rate of zero. Fortunately for us, and for the continuance of evolution, this genetic nirvana is never quite attained. Natural selection, the second stage in the Darwinian process, is a non-random force, pushing towards improvement. Mutation, the first stage in the process, is random in the sense of not pushing towards improvement. All improvement is therefore, in the first place, lucky, which is why people mistakenly think of Darwinism as a theory of chance. But mistaken they are.

The belief that natural selection favours a mutation rate of zero and that mutation is undirected does not preclude an intriguing possibility, which I have called 'the evolution of evolvability', and advocated in an essay of that title. I'll explain a new version of the idea—kaleidoscopic embryology—in Chapter 7. Meanwhile, let us return to natural selection itself, the other half of the Darwinian partnership. Though mutation is allowed to be random and in one important sense almost certainly is random, the whole essence of natural selection is that it is *not* random. Of all the wolves that might survive, a non-random sample—the fleetest of foot, the canniest of wit, the sharpest of sense and tooth—are the ones that do survive and pass on their genes. Consequently, the genes that we see in the present are copies of a non-random sample of the genes that have existed in the past. Every generation is a gene sieve. The genes that still exist after a million generations of sieving have what it takes to get through sieves. They have participated in the embryonic constructing of a million bodies without a single failure. Every one of those million bodies has survived to adulthood. Not one of them was too unattractive to find a mate—unattractive meaning whatever seems unattractive to would-be mates of the species concerned. Every single one of them proved capable of bearing or begetting at least one child. The sieve is an exacting one. The genes that penetrate to the future are not a random sample, they are an élite. They have survived ice ages and droughts, plagues and predators, busts and booms of population. They have survived shifting climates not just in the conventional sense of rain, ice and drought. They have survived shifting climates of companion genes, for the lot of a gene where there is sexual reproduction is to change partners in every generation; surviving genes are those that flourish when rubbing shoulders with successive samplings from the genes of the whole species, and this means other genes that are good at cooperating with the other genes of the species. The dominant part of the climate in which a gene has to survive is the other genes of the species: its companions in the 'River out of Eden' that flows through successive bodies down the generations. We can think of different species, as they split apart in branchings of the river, as separate micro-climates in which different sets of genes have to survive.

For simplicity we speak of mutation as the first stage in the Darwinian process, natural selection as the second stage. But this is misleading if it suggests that natural selection hangs about waiting for a mutation which is then either rejected or snapped up and the waiting begins again. It could have been like that: natural selection of that kind would probably work, and maybe does work somewhere in the universe. But as a matter of fact on this planet it usually isn't like that. There is actually a large pool of variation, originally fed by a trickle of mutations but importantly stirred and agitated into greater variation by sexual reproduction. Variation comes originally from mutation but the mutations may be quite old by the time natural selection gets around to working on them.

For instance, my Oxford colleague the late Bernard Kettlewell famously studied the evolution of dark, almost black, moths in species that had hitherto been light-coloured. In the species he especially studied, *Biston betularia,* dark individuals tend to be slightly hardier than light ones, but in unpolluted country districts they are rare because they are conspicuous to birds and are promptly picked off. In industrial areas where tree trunks have been blackened by pollution, they are less conspicuous than the light-coloured forms and consequently less likely to be eaten. This also allows them to enjoy the additional advantage of their natural hardiness. The consequent increase in numbers of dark forms, to overwhelming numerical dominance in industrial areas since the mid-nineteenth century, has been dramatic and is one of the best attested examples of natural selection in action. And now we come to the reason for introducing the case here. It is often wrongly thought that after the Industrial Revolution natural selection worked on a single brand-new mutation. On the contrary, we can be sure that there have always been dark individuals—they just haven't lasted very long. Like most mutations, this one will have been recurrent but the dark moths were always rapidly snapped up by birds. When conditions changed after the Industrial Revolution, natural selection found a ready-made minority of dark genes in the gene pool to work upon.

We have identified the ingredients that must be present before evolution can occur as being mutation and natural selection. These two will follow automatically on any planet given a more fundamental ingredient, one that is difficult, but obviously not impossible, to procure. This difficult basic ingredient is heredity. In order for natural selection to occur, anywhere in the universe, there must be lineages of things that resemble their immediate ancestors more than they resemble members of the population at large. Heredity is not the same thing as reproduction. You can have reproduction without heredity. Bush fires reproduce but without heredity.

Imagine a dry, parched grassland, stretching for mile after mile in every direction. Now, in a particular place, a careless smoker drops a lighted match and in no time at all the grass has flared up into a racing fire. Our smoker runs away from it as hard as his coughing lungs will let him, but we are more concerned with the way the fire spreads. It doesn't just swell steadily outwards from the original starting point. It also sends sparks up into the air. These sparks, or burning wisps of dry grass, are carried by the wind far away from the original fire. When a spark eventually comes down, it starts a new fire somewhere else on the tinder-dry prairie. And later, the new fire sends off sparks which kindle yet more new fires somewhere else. We could say that the fires are indulging in a form of reproduction. Each new fire has one parent fire. This is the fire that spat out the spark that started it. And it has one grandparent fire, one great-grandparent fire, and so on back until the ancestral fire started by the wayward match. A new fire can have only one mother but it can have more than one daughter, because it can send out more than one spark in different directions. If you watched the whole process from above and were able to record the history of each flare-up, you could draw out a complete family tree of the fires on the prairie.

Now the point of the story is that although there is reproduction among the fires there is no true *heredity*. For there to be true heredity, each fire would have to resemble its parent more than it resembled the other fires in general. There is nothing wrong with the *idea* of a fire resembling its parent. It could happen. Fires do vary, do have individual qualities, just as people do. A fire may have its own characteristic flame colour, its own smoke colour, flame size, noise level and so on. It could resemble its parent fire in any of these characteristics. If, in general, fires *did* resemble their parents in these ways, we could say that we had true heredity. But in fact fires don't resemble their parents any more than they resemble the general run of fires dotted around the prairie. An individual flare-up gets its characteristic qualities, its flame size, smoke colour, crackle volume and so on, from its surroundings; from the kind of grass that happens to be growing where the spark lands; from the dryness of the grass; from the speed and direction of the wind. These are all qualities of the local area where the spark lands. They are not qualities of the parent fire from which the spark came.

In order for there to be true heredity, each spark would have to carry with it some quality, some characteristic essence, of its parent fire. For example, suppose that some fires have yellow flames, others red flames, others blue. Now, if yellow-flamed fires give off sparks that start new yellow-flamed fires, whereas red-flamed fires give off sparks that start new red fires, and so on, we'd have true heredity. But that isn't what happens. If we see a blue flame we say, 'There must be some copper salts in this area.' We don't say, 'This fire must have been started by a spark from another blue-flamed fire somewhere else.'

And this, of course, is where rabbits and humans and dandelions differ from fires. Don't be misled, incidentally, by the fact that rabbits have two parents and four grandparents whereas fires have only one parent and one grandparent. That is an important difference, but it is not the one we are talking about at the moment. If it helps, think not of rabbits but of stick insects or aphids, where females can have daughters, granddaughters and great-granddaughters without males ever being involved. The shape, colour, size and temperament of a stick insect is influenced, no doubt, by the place and climate of its upbringing. But it is influenced, too, by the spark that flies only from parent to child.

So what is it, this mysterious spark that flies from parent to offspring but not from fire to fire? On this planet it is DNA. The most amazing molecule in the world. It is easy to think of DNA as the information by which a body makes another body like itself. It would be more correct to see a body as the vehicle used by DNA to make more DNA like itself. All the DNA in the world at a given time, such as now, has come down through an unbroken chain of successful ancestors. No two individuals (except identical twins) have exactly the same DNA. Differences between DNA in individuals really contribute to their survival and chances of reproducing that same DNA. To repeat because it is so important, the DNA that has made it down the river of time is DNA that has, for hundreds of millions of years, inhabited the bodies of successful ancestors. Lots of would-be ancestors have died young, or failed to find a mate. But none of their DNA is still with us in the world.

It would be easy, at this point, to make the mistake of thinking that something, some elixir of success, some odour of sanctity from the good, successful, ancestral bodies 'rubs off' on the DNA as it passes through them. Nothing of the kind occurs. The river of DNA that flows through us into the future is a pure river that (mutations apart) leaves us exactly as it finds us. To be sure, it is continually mixed in sexual recombination. Half the DNA in you is from your father and the other half is from your mother. Each one of your sperms or eggs will contain a different mixture assembled from the genetic streamlet that came from your father and the genetic streamlet that came from your mother. But the point I was making remains true. Nothing about successful ancestors 'rubs off' on the genes as they pass through on their way to the distant future.

The Darwinian explanation for why living things are so good at doing what they do is very simple. They are good because of the accumulated wisdom of their ancestors. But it is not wisdom that they have learned or acquired. It is wisdom that they chanced upon by lucky random mutations, wisdom that was then selectively, non-randomly, recorded in the genetic database of the species. In each generation the amount of luck was not very great: small enough to be believable even by the sceptical physicists whom I quoted earlier. But, because the luck has been accumulated over so many generations, we are eventually very impressed by the apparent improbability of the end product. The whole Darwinian circus depends upon—follows from—the existence of heredity. When I called heredity the basic ingredient, I meant that Darwinism, and hence life, will follow, more or less inevitably, on any planet in the universe where something equivalent to heredity arises.

We have arrived back at Mount Improbable, back to 'smearing out' the luck: to taking what looks like an immense amount of luck—the luck needed to make an eye where previously there was no eye, say—and explaining it by splitting it up into lots of little pieces of luck, each one added cumulatively to what has gone before. We have now seen how this actually works, by means of the accumulation of lots of little pieces of ancestral luck in the DNA that survives. Alongside the minority of genetically well-endowed individuals who survived, there were large numbers of less favoured individuals who perished. Every generation has its Darwinian failures but every individual is descended only from previous generations' successful minorities.

The message from the mountain is threefold. First is the message we have already introduced: there can be no sudden leaps upward—no precipitous increases in ordered complexity. Second, there can be no going downhill—species can't get worse as a prelude to getting better. Third, there may be more than one peak—more than one way of solving the same problem, all flourishing in the world.

Take any part of any animal or plant, and it is a sensible question to ask how that part has been formed by gradual transformation from some other part of an earlier ancestor. Occasionally we can follow the process through successively younger fossils. A famous example is the gradual derivation of our mammalian ear bones—the three bones that relay sound (with exquisite impedance-matching, if you should happen to know the technical jargon) from the ear-drum to the inner ear. Fossil evidence clearly shows that these three bones, called the hammer, the anvil and the stirrup, are lineally descended from three corresponding bones that, in our reptile ancestors, formed the jaw joint.

Often the fossil record is less kind to us. We have to guess at possible intermediates, sometimes with a bit of inspiration from other modern animals which may or may not be related. Elephant trunks contain no bones and do not fossilize, but we don't need fossils to realize that the elephant's trunk started out as just a nose. It is now . . . well, let me quote from a book that has me abashedly forcing back the tears whenever I read it: *Battle for the Elephants,* by a couple of heroes, Iain and Oria Douglas-Hamilton. They wrote alternate chapters and here is Oria's horrified description, on page 220, of an elephant 'cull' that she witnessed in Zimbabwe:

> I looked at one of the discarded trunks and wondered how many millions of years it must have taken to create such a miracle of evolution. Equipped with fifty thousand muscles and controlled by a brain to match such complexity, it can wrench and push with tonnes of force. Yet, at the same time, it is capable of performing the most delicate operations such as plucking a small seed-pod to pop in the mouth. This versatile organ is a siphon capable of holding four litres of water to be drunk or sprayed over the body, as an extended finger and as a trumpet or loud speaker.
>
> The trunk has social functions, too; caresses, sexual advances, reassurances, greetings and mutually intertwining hugs; and among males it can become a weapon for beating and grappling like wrestlers when tusks clash and each bull seeks to dominate in play or in earnest. And yet there it lay, amputated like so many elephant trunks I had seen all over Africa.

The paragraph has had the usual effect on me . . .

Here, the message from the mountain is that the ancestors of elephants must have included a continuous series of intermediates with more or less longish noses like tapirs, or elephant shrews, or proboscis monkeys, or elephant seals. None of these creatures is closely related to elephants (or to each other). All have evolved their long noses independently of each other and probably for different reasons (Figure 3.1).

In the evolution of the elephant from its short-nosed ancestors, there must have been a smooth, gradual succession of steadily longer noses, a sliding gradient of thickening muscles and more intricately dissected nerves. It must have been the case that, as each extra inch was added to the length of the average trunk, the trunk became better at its job. It must never be possible to say anything like: 'That me-

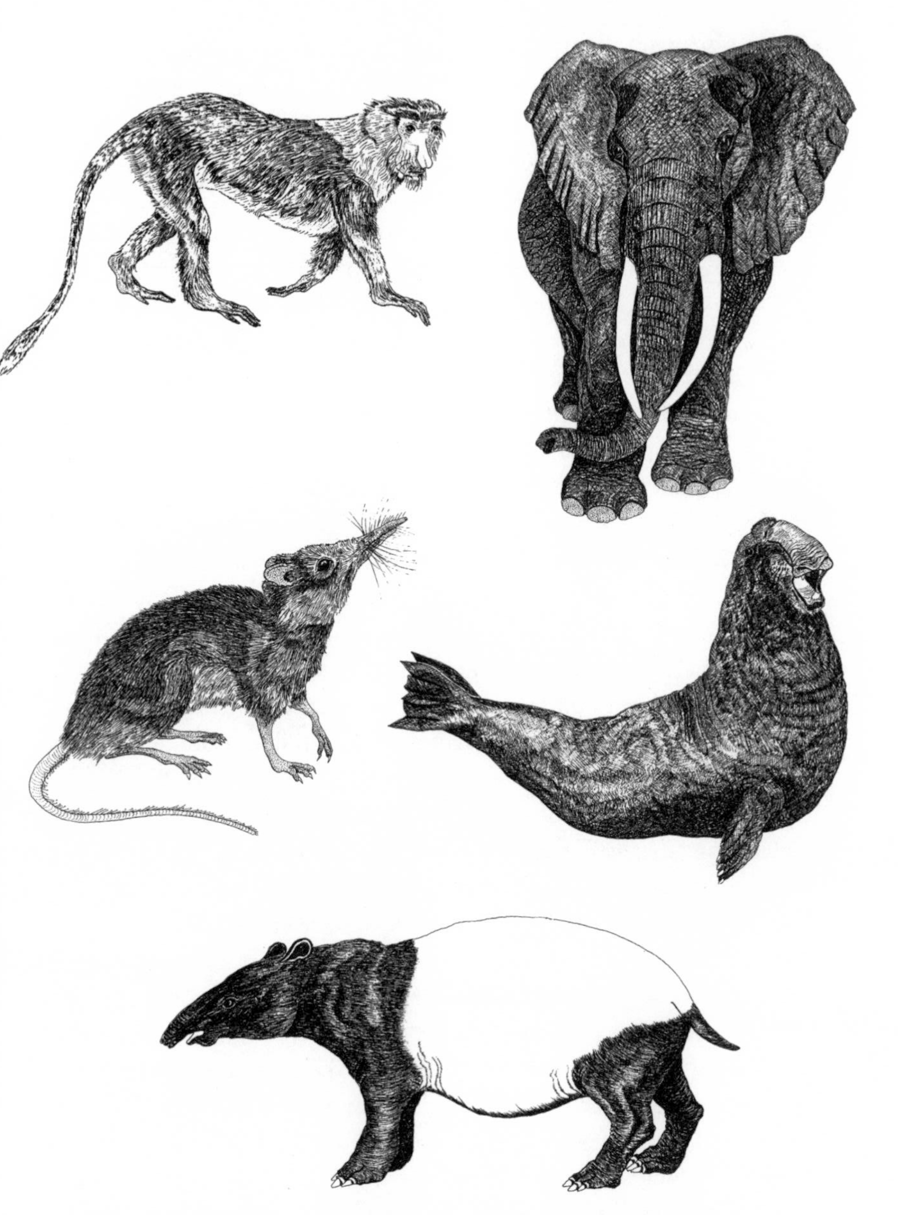

Figure 3.1 African elephant, *Loxodonta africana,* and unrelated long-nosed mammals who probably developed their long noses for independent reasons: (anticlockwise from top left) proboscis monkey, *Nasalis larvatus*; elephant shrew, *Rhynchocyon petersi;* Malayan tapir, *Tapirus indicus*; and southern elephant seal, *Mirounga leonina.*

dium-sized trunk is no good because it is neither one thing nor the other—falls between two stools—but don't worry, give it another few million years and it'll be fine.' No animal ever made a living purely by being on the evolutionary path to something better. Animals make a living by eating, avoiding being eaten, and reproducing. If a medium-sized trunk were always less efficient for these purposes than either a small nose or a big trunk, the big trunk would never have evolved.

Just because the trunk had to be useful in all its intermediate stages, this does not mean that it had to be useful for the same purpose throughout all the intermediate stages. Early elongation could have provided an advantage that had nothing to do with picking up objects. Perhaps the nose got longer in the first place to jack up the sense of smell, as in elephant shrews; or as a resonator for calls, as in elephant seals; or an adornment to attract a mate, as in—however surprising it may seem to our aesthetic sense—proboscis monkeys. On the other hand it is also possible that its usefulness as a 'hand' came quite early in elephant evolution, when it was still quite short. This guess is made plausible by comparison with tapirs who use their nose to grasp leaves and pull them towards the mouth. Independent evolutions of similar devices in different animals can illuminate our understanding of each other.

In the specific case of the elephant trunk, there is suggestive evidence from the hard parts of the skull that fossilize, especially the tusks and associated bones. Today's two species of elephants are the sole survivors of a once rich radiation of tusked animals thriving in every continent. Modern elephant tusks are the enormously enlarged upper incisor teeth but many fossil forms, such as some of the mastodonts, had more prominent lower incisors also pointing forwards. Sometimes they were large and spiky like the tusks that we now see only in the upper jaw. In other kinds they were flat so that the two great teeth together constituted a broad shovel or spade of ivory prolonging the line of the lower jaw, which was probably used for grubbing up roots and tubers. The shovel extended so far in front of the lower jaw that the upper lip could not reach the food that it dug up. It seems probable that the incipient trunk extended originally to work against the shovel and grasp the food that the shovel dug. Later, we may guess, the nascent trunk became so good at this that it started to be used on its own, without the shovel. Later, at least in the lineages that have survived, the shovel itself became reduced while the trunk remained, as if beached by the receding tide. The lower jaw retreated to something like its original proportions, leaving the now fully independent trunk as its legacy. Consult John Maynard Smith's excellent *The Theory of Evolution*, pages 291–4, for a fuller account of elephant-trunk evolution.

The word 'pre-adaptation' is used for cases where an organ is originally used for some purpose and then later in evolution is taken over for another purpose. It is an illuminating idea, for it frequently rescues us from puzzlement over evolutionary origins. Porcupine quills are now formidable weapons. They didn't spring from nothing, but are modified hairs, 'pre-adapted' for the completely different purpose of keeping warm. Many mammals have highly developed, specialist scent glands. Where they sprang from might seem a mystery until you look at them closely under the microscope and see that they are modified from a smaller gland with the very different purpose of secreting sweat to cool the body down. Unreconstructed sweat glands are still common in other parts of the very same animal, so the comparison is easy to make. Other scent glands seem to have evolved from sebaceous glands, whose original duty was to protect hairs with a waxy secretion. Often the pre-adaptation and its modern successor are not unrelated. Sweat happens to smell, and it happens to be secreted when the animal is emotionally aroused (people are popularly supposed to sweat with fear, and I know I do when an important lecture isn't going according to plan). It therefore was natural for the old pre-adaptation to shift into its specialized counterpart.

Sometimes it isn't obvious which came first—which is the early pre-adaptation and which the later specialization. Darwin, wondering about the evolutionary origin of the lung, sought an answer in the swim bladder of fish. This is a gas-filled bladder, which bony fish use to control their buoyancy on the principle of the Cartesian Diver (those little men in bottles who can be made to sink or rise by means of gentle pressure on the cork). By using muscles to adjust the volume of its swim bladder, a fish is able to change the depth at which it can rest at equilibrium.

This applies only to ordinary bony fish. Sharks (which, despite their fishy shape, are actually more distantly related to bony fish than we are) lack swim bladders and they consequently have to do more swimming work to keep themselves at their desired level in the water. The swim bladder looks like a lung, and Darwin thought that it might be the pre-adaptation from which our lungs evolved. Modern zoologists mostly reverse this particular horse and cart, suspecting that the swim bladder is a recent modification of a primitive lung (air-breathing fish are quite common, to this day). Whichever is the more primitive, we need to think about what, even earlier, preceded it. Perhaps the lung/swim bladder arose from a pouch of the gut and had a primitive digestive function. In every stage of its evolution, at every step up the slopes of Mount Improbable, the pouch/cavity/lung had to be useful to the animal that bore it.

Couldn't the elephant's trunk have shot out in a single, giant step? Why shouldn't an offspring have had a trunk like an elephant, while its parents had trunks like tapirs? There are really three questions here. First, the question whether mutations of very large magnitude—macro-mutations—can happen. Second, the question whether, given that they happen, natural selection would ever favour them. Third is a more subtle question about what we mean by large, when we speak of large mutational change. I shall come back to a distinction I have made in an earlier book, between 'Boeing 747 macro-mutations' and 'Stretched DC8 macro-mutations'.

The answer to the first of the three question is yes. Macro-mutations do happen. Offspring are sometimes born radically, monstrously different from either parent, and from other members of the species. The toad in Figure 3.2 is said by the photographer, Scott Gardner of the *Hamilton Spectator,* to have been found by two girls in their garden in Hamilton, Ontario. He says that they put it on the kitchen table for him to photograph. It had no eyes at all on the outside of its head. When it opened its mouth, Mr Gardner said, it seemed to become more aware of its surroundings. He said that it was taken for examination to the Veterinary Department of Guelph University, but I have not so far discovered any full report on it. Such unfortunate freaks are interesting because they often give us clues

Figure 3.2 Macro-mutations do happen. This freak toad with eyes in the roof of its mouth is said to have been found surviving wild in a Canadian garden. This photograph was originally published in a local newspaper, *The Hamilton Spectator.*

about how embryonic development happens normally. Not all human birth defects are genetic, for instance those caused by thalidomide, but many are. A simple dominant gene causes achondroplasia, a severe shortening of the limb bones resulting in low stature and unusual proportions. Mutations of large effect like this—'macro-mutations'—are sometimes called saltations. The achondroplasia gene is usually inherited from one parent, but it very occasionally springs up spontaneously by mutation, and this is how it must arise originally. A similarly dramatic mutation could, in theory though I very much doubt it in practice, have given rise to an abrupt and sudden nose extension from tapir length to elephant length in a single generation.

Coming to the second question, of whether, once a large 'freak' macro-mutation had arisen, natural selection would ever favour it, you might think that it is not the sort of question that has a general answer. Doesn't it vary among particular cases, say, yes for achondroplasia, no for two-headed calves? The dog equivalent of the achondroplasia gene has in fact been positively favoured in artificial selection by human breeders, not just to service idle whims but to produce useful, working dogs. Dachshunds were bred to go down badger sets, and a significant part of the genetic sculpting that led to the breed was the incorporation of the achondroplasia gene. Perhaps it sometimes happens in nature that major mutations, like achondroplasia, suddenly open up a new way of life or a new diet: a dwarf animal, although heavily penalized when pursuing prey over open country, suddenly discovers that, unlike most of its colleagues, it can follow the prey down a hole.

Evolutionary theorists have sometimes suggested that major saltations are incorporated into evolutionary change in nature. The famous German-American geneticist Richard Goldschmidt advocated the theory under the memorable catchphrase of the 'hopeful monster' theory. I'll mention one possible example in Chapter 7. But Goldschmidt's theory has never been widely supported, and there are general reasons for doubting whether macro-mutations or freaks really are important in evolution. Organisms are extremely complicated and sensitively adjusted pieces of machinery. If you take a complicated piece of machinery, even one which is not working all that well, and make a very large, random alteration to its insides, the chance that you will improve it is very low indeed. If, on the other hand, you make a very small random alteration to its insides, you may improve it. If your television aerial (antenna) is not quite properly aligned, a very slight random twist to the aerial has about a 50–50 chance of improving matters. This is because, whichever direction the aerial ought to be pointing in, there's a 50 per cent chance that your slight random twist will be in that direction. But a very large random assault on the aerial, wrenching it round violently through a very large angle, is more likely to make matters worse. This is partly because, even if your twist is in the right direction, it will probably overshoot the correct angle. More generally it is because there are so many more ways of being wrongly adjusted than of being rightly adjusted. A complicated mechanism that is working at all cannot be too far from correct adjustment. A small random change may improve it; or, if it makes matters worse, it will still not move it too far from the correct arrangement. But a very large random change has the effect of sampling the gigantic set of all possible arrangements. And the vast majority of all possible arrangements are wrong.

Even the common experience that a machine that is on the blink can often be improved by a good kick does not contradict my argument. Violent though the kick may be, the television set is a robust piece of hardware and the kick doesn't necessarily have a large effect on the arrangement of the parts. What it can do is slightly change the position of any part that is slightly loose; and this loose part is quite likely to be the very part that is causing the faulty behaviour.*

Turning to living creatures, I wrote in *The Blind Watchmaker* that *however many ways there may be of being alive, it is certain that there are vastly more ways of being dead.* (I'd be inhuman not to confess my delight that this remark has made it into the *Oxford Dictionary of Quotations!*) If you think of all possible ways of arranging the bits of an animal, almost all of them would turn out to be dead; more accurately they'd mostly never be born. Each species of animal and plant is an island of workability set in a vast sea of conceivable arrangements most of which would, if they ever came into existence, die. The ocean of all possible animals includes animals with their eyes in the soles of their feet, animals with lenses in their ears instead of their eyes, animals with one left wing and one right fin; animals with skulls around their stomachs and nothing around their brains. There is no point in going on inventing. I have said enough to demonstrate that the islands of survivability, however large and however numerous they may be, are

*Judith Flanders has called my attention to the following amusingly relevant story in Robert X. Cringely's book, *Accidental Empires.* The story concerns the Apple III, a desktop computer of the generation between the famous Apple II and the even more famous Macintosh, launched in 1980: '. . . the automated machinery that inserted dozens of computer chips on the main circuit board didn't push them into their sockets firmly enough. Apple's answer was to tell 90,000 customers to pick up their Apple III carefully, hold it twelve to eighteen inches above a level surface, and then drop it, hoping that the resulting crash would reseat all the chips.'

minuscule in size and infinitesimal in number when compared with the ocean of dead unworkability.

When a parent has a mutant child, the parent, being alive, must be safely ensconced on one of the islands. A small mutation—a fractional lengthening of a leg-bone here, a delicate adjustment to a jaw angle there—simply moves the child to a different part of the same island. Or it may reclaim a small offshore sandbank and join it to the dry land. But a large mutation, a drastic, freakish, revolutionary change, is equivalent to a mad leap into the wild blue yonder. The macro-mutant is catapulted in a random direction, leagues away from its home island. It is just *possible* that it will chance to land on another island. But since islands are so few and small, and the sea so large, the chances are very very low. It may happen very occasionally once every few million years, and when it does happen it may have a dramatic impact on the course of evolution.

We mustn't push the metaphor of the islands too far. There is much wrong with it. All species are related to each other, which means there must be ways of travelling, through the ocean of possibilities, from any one way of being alive to every other. The metaphor of the islands does not help us here, and the metaphor of Mount Improbable is better. The islands serve the particular purpose of dramatizing the point that the more drastic and freakish the mutation the less likely it is to be favoured.

We also need to distinguish different kinds of macro-mutation. By invoking imaginary animals with eyes in their soles and lenses in their ears, I concentrated attention on changes in arrangement of parts. Large changes of this kind are certainly very unlikely indeed to strike lucky and survive. But there are also large changes in magnitude of a part, which do not involve rearrangements of parts. A sudden shooting out of a tapir-like nose to an elephant-like trunk would be an example, if lengthening were all that happened. It is less obvious that a drastic change of this kind necessarily constitutes a leap into the ocean of impracticality or death.

I promised that I'd return to 'Boeing 747' and 'Stretched DC8' macro-mutations. Remember Sir Fred Hoyle's debating point about junkyards and 747s? He is reported to have said that the evolution, by natural selection, of a complicated structure such as a protein molecule (or, by implication, an eye or a heart) is about as likely as a hurricane's having the luck to put together a Boeing 747 when whirling through a junkyard. If he'd said 'chance' instead of 'natural selection' he'd have been right. Indeed, I regretted having to expose him as one of the many toilers under the profound misapprehension that natural selection *is* chance. Any theory that expects evolution to put together a new, complex machine like an eye or a haemoglobin molecule, in a single step from nothing, is asking too much of chance. On this theory, natural selection has hardly any work to do. All the 'design' work is being put in by mutation, a single large mutation. It is this kind of macro-mutation that deserves the metaphor of the 747 and the junkyard, and I call it a Boeing 747 macro-mutation. Boeing 747 macro-mutations do not exist and they have no connection with Darwinism.

Turning to my other airliner analogy, the Stretched DC8 is like an ordinary DC8 only rather longer. The fundamental design of the DC8 is all there, but an extra length of tubing has been let into the middle of the fuselage. There are also more seats, more luggage lockers and more of all the other things that repeat down the length of a plane. Equally obviously there are extra lengths of the cables, tubes and carpets that run the length of an airliner's fuselage. Slightly less obviously, there surely must be numerous consequential modifications to other parts of the plane, necessitated by the new task of lifting a greater length of the fuselage off the ground. But, fundamentally, the difference between the DC8 and the Stretched DC8 comes down to a single macro-mutation: the fuselage is abruptly and suddenly much longer than its predecessor. There was not a gradual series of intermediates.

Giraffes have evolved from an ancestor rather like a modern okapi (Figure 3.3). The most conspicuous change is the elongation of the neck. Could this have come about in a single, large mutation? I hasten to say that I am sure it didn't. But that is another matter from saying that it couldn't. A Boeing 747 mutation like a brand-new complex

Figure 3.3 Steps to a long neck. Okapi, *Okapia johnstoni,* which may be similar to an ancestor of modern giraffes, with giraffe, *Giraffa camelopardalis reticulata.*

eye—complete with iris diaphragm and refocusable lens, springing from nothing, like Pallas Athene from the brow of Zeus—that can *never* happen, not in a billion billion years. But, like the stretching of the DC8, the giraffe's neck could have sprung out in a single mutational step (though I bet it didn't). What is the difference? It isn't that the neck is noticeably less complicated than the eye. For all I know it may be more complicated. No, what matters is the complexity of the *difference* between the earlier neck and the later one. This difference is slight, at least when compared with the difference between no eye and a modern eye. The giraffe's neck has the same complicated arrangement of parts as the okapi (and presumably as the giraffe's own short-necked ancestor). There is the same sequence of seven vertebrae, each with its associated blood vessels, nerves, ligaments and blocks of muscle. The difference is that each vertebra is a lot longer, and all its associated parts are stretched or spaced out in proportion.

The point is that you may only have to change one thing in the developing embryo in order to quadruple the length of the neck. Say you just have to change the rate at which the vertebral primordia grow, and everything else follows. But in order to make an eye develop from bare skin you have to change, not one rate but hundreds (see Chapter 5). If an okapi mutated to produce a giraffe's neck it would be a Stretched DC8 macro-mutation, not a 747 macro-mutation. It is therefore a possibility which need not be totally ruled out. Nothing new is added, in the way of complication. The fuselage is elongated, with all that that entails, but it is a stretching of existing complexity, not an introduction of new complexity. The same would be true even if the giraffe had more than seven segments in its neck. The number of vertebrae in different species of snakes varies from about 200 to 350. Since all snakes are cousins of each other, and since vertebrae cannot come in halves or quarters, this must mean that, from time to time, a snake is born with at least one more, or one fewer, vertebra than its parents. These mutations deserve to be called macro-mutations, and they have evidently been incorporated in evolution because all these snakes exist. They are DC8 mutations because they involve the duplication of existing complexity, not the 747 invention of new complexity.

There is something that could come to the evolutionary aid of the freak macro-mutant, namely the fact that the effect of a given gene depends upon the other genes that are present in the same body. The effect of a gene on a body, its so-called phenotypic effect, is not written on its surface. There is nothing in the DNA code of the achondroplasia gene that a molecular biologist could decode as 'short' or 'dwarf'. It has the effect of making limbs short only when surrounded by lots of other genes, to say nothing of other features of the environment. A gene's meaning is context-dependent. The embryo develops in a climate produced by all the genes. The effect that any one gene has on the embryo depends upon the rest of the climate. R. A. Fisher, whom I've already quoted, expressed this long ago by saying that some genes act as 'modifiers' of the effects of others. Notice that this doesn't mean genes modify the DNA code of others. Certainly not. Modifiers simply change the climate in such a way as to modify the effects of other genes on bodies, not the DNA sequence of the other genes.

As we have seen, it is not (quite) totally inconceivable that a parent with a six-inch, tapir-like proboscis could have produced a mutant child with a five-foot elephant-like trunk in a single generation, due to a one-gene change—a macro-mutation. It is very unlikely that the new nose would immediately behave like a good, working trunk. This is where 'modifier' genes, and the notion of a 'climate' of other genes, could theoretically come to the rescue. So long as the macro-mutation is at least approximately good for something, so that individuals possessing it don't die out, subsequent selection of modifier genes can clean up the details and smooth off the rough corners. Think of the advent into the population of the major mutation as equivalent to a cataclysmic challenge, like an ice age. Just as a new ice age causes a whole collection of genes to be selected, so also does a drastic mutational change to the average body, such as a sudden elongation of the nose.

Genes that 'clean up' in the wake of a new major mutation do not work only on the major gene's most obvious effects. They may act on some unexpectedly distant part of the body to compensate for, mitigate the ill-effects of, or enhance the possible benefits of, a major mutation. In the wake of a greatly enlarged nose, since the trunk increases the weight of the head, the bones of the neck will need to be strengthened. The balance of the whole body may change, with further knock-on effects, perhaps on the backbone and the pelvis. All this consequential selection works upon dozens of genes affecting many different parts of the body.

Although I introduced the idea of 'cleaning up afterwards' in the context of major macro-mutations, this kind of selection is certainly important in evolution whether or not there are any macro-mutational steps. Even micro-mutations arouse consequences such that 'cleaning up afterwards' is highly desirable. Any gene can act as a modifier of the effects of any other. Many genes modify each other's effects. Some authorities would go as far as to say that, of those genes that have any effect at all (lots don't), most genes modify most other genes' effects. This is another aspect of what I meant when I said that the 'climate' in which a gene has to survive consists largely of the other genes of the species.

At the risk of spending longer on macro-mutations than they deserve, there remains one possible source of confusion that I must anticipate. There is an expertly publicized, and not uninteresting, theory known as 'punctuated equilibrium'. To go into detail would take me beyond this book's scope. But, because the theory is heavily promoted and widely misunderstood, I must just stress that the theory of punctuated equilibrium does not have—or should not be represented as having—any legitimate connection with macro-mutation. The theory proposes that lineages spend long periods in stasis, undergoing no evolutionary change, punctuated by occasional rapid bursts of evolutionary change which coincide with the birth of a new species. But, rapid though these bursts may be, they are still spread over large numbers of generations, and they are still *gradual*. It is just that the intermediates usually pass too quickly to be recorded as fossils. This 'punctuation as rapid gradualism' is very different from macro-mutation, which is instantaneous change in a single generation. The confusion arises partly because one of the two advocates of the theory, Stephen Gould (the other is Niles Eldredge), also independently happens to have a soft spot for certain kinds of macro-mutations, and he occasionally underplays the distinction between rapid gradualism and true macro-mutation—not, I hasten to add, miraculous Boeing 747 macromutation. Eldredge and Gould are rightly annoyed at the misuse of their ideas by creationists who, in my terminology, think that punctuated equilibrium is about huge, 747-type macro-mutations which, they are right to believe, would require miracles. Gould says:

> Since we proposed punctuated equilibria to explain trends, it is infuriating to be quoted again and again—whether through design or stupidity, I do not know—as admitting that the fossil record includes no transitional forms. Transitional forms are generally lacking at the species level, but they are abundant between larger groups.

Dr Gould would lessen the risk of such misunderstanding if he more clearly emphasized the radical distinction between rapid gradualism and saltation (i.e. macro-mutation). Depending upon your definition, the theory of punctuated equilibrium is either modest and possibly true or it is revolutionary and probably false. If you blur the distinction between rapid gradualism and saltation you may make the punctuation theory seem more radical. But at the same time you offer an open invitation to misunderstanding, an invitation that creationists are not slow to take up.

There is a supremely banal reason why transitional forms are generally lacking at the species level. I can explain it best with an analogy. Children turn gradually and continuously into adults but, for legal purposes, the age of majority is taken to be a particular birthday, often the eighteenth. It would therefore be possible to say, 'There are 55 million people in Britain but not a single one of them is intermediate between non-voter and voter.' Just as, for legal purposes, a juvenile changes into a voter as midnight strikes on the eighteenth birthday, so zoologists always insist on classifying a specimen as in one species or another. If a specimen is intermediate in actual form (as many are) zoologists' legalistic conventions still force them to jump one way or the other when naming it. Therefore the creationists' claim that there are no intermediates has to be true *by definition* at the species level, but it has no implications about the real world—only implications about zoologists' naming conventions.

To look no further than our own ancestry, the transition from *Australopithecus* to *Homo habilis* to *Homo erectus* to 'archaic *Homo sapiens*' to 'modern *Homo sapiens*' is so smoothly gradual that fossil experts are continually squabbling about how to classify particular fossils. Now look at the following, from a book of anti-evolution propaganda: 'the finds have been referred to either *Australopithecus* and hence are apes, or *Homo* and hence are human. Despite more than a century of energetic excavation and intense debate the glass case reserved for mankind's hypothetical ancestor remains empty. The missing link is still missing.' One is left wondering what a fossil has to do to qualify as an intermediate. In fact the statement quoted is saying nothing whatever about the real world. It is saying something (rather dull) about naming conventions. No 'missing link', *however* precisely intermediate it was, could escape the terminological *force majeure* that would thrust it one side of the divide or the other. The proper way to look for intermediates is to forget the naming of fossils and look, instead, at their actual shape and size. When you do that, you find that the fossil record abounds in beautifully gradual transitions, although there are some gaps too—some very large and accepted, by everybody, as due to animals simply failing to fossilize. In a way, our naming procedures are set up for a pre-evolutionary age when divides were everything and we did not expect to find intermediates.

We've taken a preliminary look at Mount Improbable and seen the difference between the forbidding cliffs on one side and forgiving slopes on the other. The next two chapters look carefully at two of the peaks beloved of creationists because their cliffs seem particularly steep: first wings ('what is the use of half a wing?') and then eyes ('the eye won't work at all until all its many parts are in place, therefore it can't have evolved gradually').

CHAPTER 4

GETTING OFF THE GROUND

TO FLY HAS FOR SO LONG BEEN A HOPELESS DREAM OF humanity and, when we achieve it, we do so with such difficulty that it is easy to exaggerate how hard it is. Flying is second nature to the majority of animal species. To modify an aphorism of my colleague Robert May, to a first approximation all animal species fly. This is mostly because, as he actually said, to a first approximation all species are insects. But even if we take just warm-blooded vertebrates it is still correct to say that more than half the species fly: there are twice as many bird species as mammals, and a quarter of all mammal species are bats. Flying seems formidable to us mainly because we are large animals. There are a few larger than us like elephants and rhinos and we are naturally very aware of them but to a first approximation all animals are smaller than us (Figure 4.1).

If you are a very small animal, the conquest of the air is no problem. When you are very small, the harder challenge may be to stay on the ground. This difference, between large animals and small, follows from some inescapable principles of physics.

For objects of a given shape, weight increases disproportionately with (specifically as the third power of) length. If an ostrich egg is three times as long as a hen's egg of the same shape, it will not be three times the weight but $3 \times 3 \times 3$, or twenty-seven times the weight. Until you are used to it, this can be rather an arresting thought. If one hen's egg is breakfast for a man, one ostrich egg is breakfast for a

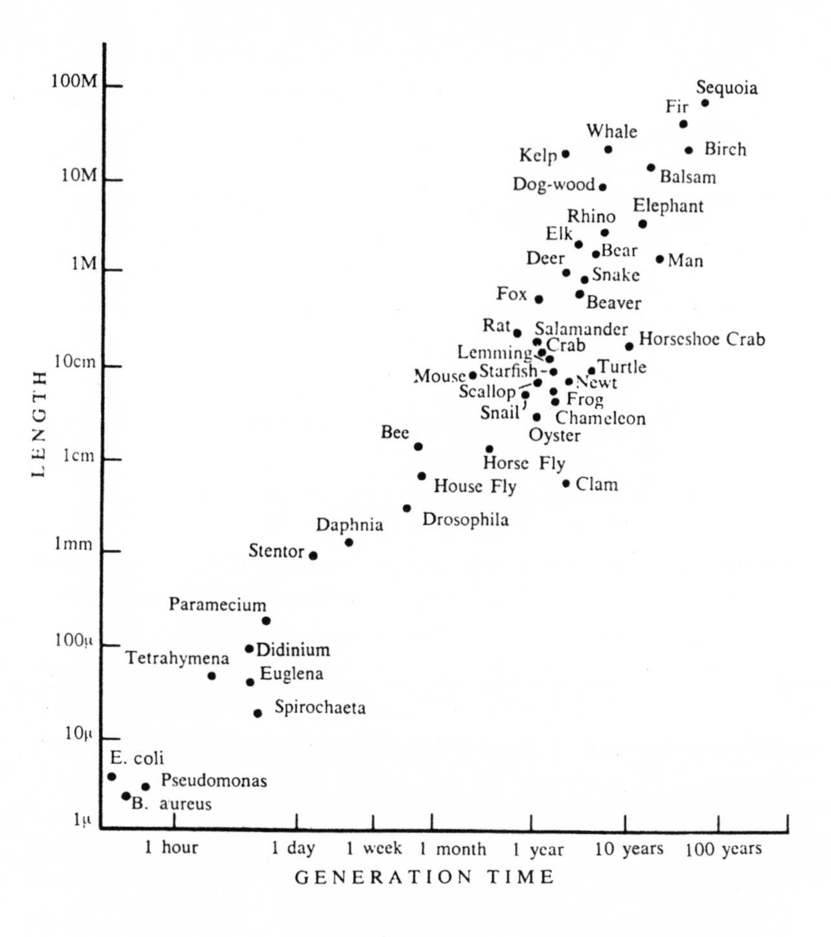

Figure 4.1 Living things vary in size over about eight orders of magnitude. To help sort the variation out, generation time is plotted against size (they are strongly correlated, for reasons not discussed here). Both axes are drawn on a logarithmic scale, otherwise paper 1,000 miles across would be required to accommodate a redwood tree on the same scale as a bacterium.

twenty-seven-man platoon. Volume, and hence weight, goes up as the third power (cube) of the linear dimension. Surface area, on the other hand, goes up as the second power (square) of the linear dimension. It is easiest to demonstrate this with cubical boxes, but the rule applies to all shapes.

Imagine a large cubical box. How many smaller boxes of exactly half the edge size could you fit in it? You can see quickly, by sketching the boxes, that the answer is eight. The big box can hold not twice as many apples as one of the smaller boxes but eight times as many apples; not twice as many tins of paint but eight times as many can be crated in the box. But if you want to paint the surface of the big box, how much more paint do you need than to paint the outside of one small box? Again, you can quickly verify by sketching the scene that the answer is neither two times nor eight times, but four times as much paint.

The difference between surface area and volume becomes more dramatic when you look at objects of very different size. Suppose a match-manufacturer, for advertising purposes, constructs a man-sized matchbox, two metres high when lying flat on the ground. A standard matchbox is two centimetres high so a stack of 100 matchboxes would just reach the height of the crate. A line of 100 matchboxes would just stretch from one end of the crate to the other. And a row of 100 matchboxes would span the width. So, if you filled the crate with matchboxes, how many would it hold? The answer is 100 × 100 × 100 or one million. In one sense the crate is only 100 times as big as an ordinary matchbox, and a naïve human eye may estimate that it is about 100 times as big. Yet in another sense it is a million times as big, and it will hold at least a million times as many matches (actually more, because relatively less space is occupied by cardboard).

If we assume that the giant matchbox is made of the same kind of cardboard as an ordinary small matchbox, what is the relative cost of the cardboard? This depends, not on the volume, nor on the linear dimensions, but on the surface area. The giant box would need, not a million times as much cardboard but a mere 10,000 times as much. The surface area of the standard, small matchbox is hugely greater *for its weight* than the surface area of the giant matchbox. If you cut up a small matchbox, you could only just stuff the folded cardboard in another small matchbox. But if you cut up our giant matchbox, the folded cardboard would hardly be noticed lining the bottom of another giant matchbox. The ratio between surface area and volume is a very important quantity. For every cubing of volume, surface area is merely squared. You can express this mathematically as the statement that, if the shape is uniformly scaled up, the ratio of surface area to volume goes up as the two-thirds power of length. The surface area to volume ratio is larger for small objects than for large objects. Small objects are more 'surfacy' than large objects of the same shape.

Now, some important things in life depend on surface area, other important things depend on volume, other important things depend on linear dimension, and yet other important things depend on various combinations of the three. Imagine a perfectly scaled-down hippopotamus, the size of a flea. The height (or length, or width) of the real hippo is perhaps a thousand times that of the flea-hippo. The weight of the hippo is then a billion times that of the flea-hippo. The surface area of the hippo is a mere million times that of the flea-hippo. So the flea-hippo has 1,000 times greater surface area *for its weight* than the large hippo. It feels like common sense to say that a scaled-down miniature hippo would find it easier to float in the breeze than a full-sized hippo, but it is sometimes important to see what lies behind common sense.

Of course big animals never are just scaled-up versions of little animals, and we can now see why. Natural selection does not allow them to be simply scaled up, because they need to compensate for such things as the change in the ratio of surface area to volume. The hippo would have about a billion times as many cells as the hippo-flea, but only about a million times as many skin cells in its outer surface. Each cell needs oxygen and food and needs to get rid of waste products, so the hippo would have about a billion times as much stuff to pass in and out of itself. The flea-hippo could use its outer skin as a significant part of the surface over which oxygen and waste products pass. But the outer skin of the full-sized hippo is, relatively, so small that it needs to increase its surface area, very substantially, in order to cope with its billion-fold greater cell population. This it does with a long folded gut, with spongy lungs and micro-tubular kidneys, the whole irrigated by a massively divided and re-divided network of blood vessels. The result is that the internal area of a large animal is spectacularly more than the area of its outer hide. The smaller an animal is, the less does it need lungs or gills or blood vessels: the outer surface of the body has a large enough area to cope, unaided, with the input–output traffic of its relatively few internal cells. A less precise way to put this is to say that a small animal has a higher proportion of its cells touching the outside world. A large animal like a hippo has such a tiny proportion of its cells touching the outside world that it has to increase that proportion with area-intensive devices like lungs, kidneys and blood capillaries.

The rate at which substances can pass in and out of a body depends upon area, but it isn't the only important thing that does. So also does the tendency to catch the air and float. The flea-hippo would be lifted by the lightest puff of wind. It could be carried high on a thermal and then float delicately back to earth where it would land softly and without injury. The real hippo, if dropped from the same height, would plummet to a terrible crash landing; and if dropped from a scaled-up proportional height it would dig its own grave. For the real hippo, flying is an impossible dream. The flea-hippo would hardly be able to help flying if it tried. To make a real hippo fly, you'd have to strap a pair of wings on it so large that . . . well, the project is doomed from the start because the mass of muscle needed to power those gigantic wings would be too heavy for the wings to lift. If you wanted to make a flying animal, you wouldn't start with a hippo.

The point is that if a large animal is going to leave the ground it has to grow large area-intensive wings, for the same kind of reason as any large animal needs surface-rich kidneys and lungs. But for a small animal to leave the ground it hardly has to grow anything at all. Its whole body is already surface-rich. There is a so-called aerial plankton, consisting of millions of small insects and other creatures floating high in the air and spreading around the world. Many of them, to be sure, have wings. But the aerial plankton also contains numerous tiny wingless creatures which float in spite of having no specialized aerofoil surfaces. They float simply because they are small, and a very tiny animal finds floating in air about as easy as we find floating in water. Indeed, the comparison goes further, for even when a tiny floating insect has wings it flaps them not so much to keep up as to 'swim' through the air. The reason 'swim' is an appropriate word is that other strange things happen when you are very small. At that scale, surface tension is such an important force that, to a small insect, the air would feel all treacly. Flapping its wings must feel, to a little insect, rather as swimming through syrup would feel to us.

You might wonder what is the use of floating without the ability to control height or steer. I won't go into the details, but dispersal *per se* can be a virtue in the eyes of genes, especially for a creature which is basically sedentary. This applies *a fortiori* to plants: any patch of ground becomes uninhabitable from time to time, say when there is a forest fire or a flood. For a plant that needs lots of sunlight, the entire forest floor is uninhabitable except at intervals when a tree falls and breaks the shade. In general, any animal or plant will be descended from ancestors that lived somewhere else and they are likely to contain genes for taking steps to disperse somewhere—anywhere—else. This is why dandelion seeds have cottony puffs. This is why burrs have hooks to stick them on animal fur. This is why many insects drift abroad in the aerial plankton and rain down on unfamiliar ground.

The ease with which small animals can float suggests that we have only to assume that flying evolved originally in small animals, and the flying peak of Mount Improbable immediately looks less formidable. Very small insects float without wings at all. Slightly larger insects are helped by tiny wing stubs to catch the breeze, and we find ourselves on a neat, shallow ramp up Mount Improbable to proper wings. Actually, it may not have been as simple as that, according to some ingenious research by Joel Kingsolver and Mimi Koehl in the University of California at Berkeley. Kingsolver and Koehl worked on the theory that the first insect wings were pre-adapted to a completely different purpose, solar panels for heating. In those early days, of course, they wouldn't have flapped. They'd have just been little projections growing out from the thorax.

Kingsolver and Koehl's research technique was cunning. They made simple wooden models, based upon the earliest known fossil insects. Some of their models had no wings. Others had wing stubs of various lengths, many of them much too short to be recognizable as wings or to fly. The model insects themselves were of a range of different sizes and they were tested in a wind tunnel to see how aerodynamically efficient they were. The models also had tiny thermometers inside them, to see how good they were at picking up artificial sunlight from a bright floodlamp.

In accordance with what we've already discussed, they found that really tiny insects float well enough with no wings at all. The slightly disconcerting result from the point of view of my simple ramp up Mount Improbable was that, at these very small sizes, small wings didn't seem to help aerodynamic efficiency. Wings didn't provide useful lift unless they were already of a substantial length. For model insects two centimetres in body length, wings as long as one body length produce significant lift. But wings of only twenty per cent of body length seem to do nothing for the animal at all. On the face of it, this looks like a precipice on Mount Improbable, because it seems to call for a single large mutation to jump the wings out to a substantial length, all in one go. It isn't a very formidable precipice, however, because of the following pair of additional facts.

In the first place, it is only for very small insects that you need relatively large wing stubs to get any aerodynamic benefit. If the insect is rather larger, tiny wing stubs *do* give some significant lift. At ten centimetres of body length, as you increase the wing stubs gradually from nothing, there is an immediate leap in aerodynamic benefit.

For our second supplementary fact, we revert to very small insect models. Here tiny wing stubs did prove to have an immediate *thermal* benefit. When tiny wings become slightly less tiny, they do not provide extra lift but they do become better solar panels. It seems that there is a smooth gradient of improvement in solar-panel performance when the insect body is very small. One millimetre stubs are better than none at all, two millimetre stubs are better than one, and so on. But the 'and so on' doesn't go on for ever. Beyond a certain length, further improvement in solar-panel performance tails off. It can be argued, therefore, that the solar-panel improvement gradient, on its own, couldn't get the stubs out to lengths where the aerodynamic function could take over. But Kingsolver and Koehl have a good solution to this. Once the stubs had evolved in small insects because of their solar-panel advantage, some insects evolved larger body size for a quite different reason. There could be any number of reasons: it is very common for animals to evolve larger size as time goes by. Perhaps larger insects have an advantage because they are less likely to be eaten. If they grew, over evolutionary time, for whatever reason, it can be presumed that their solar-panel stubs would have grown with them, automatically. Now, a consequence of this general size increase is that the insects, stubs and all, would automatically be carried into the size range where aerodynamic benefits could take over and continue the steady push up Mount Improbable, albeit up a different slope towards a different peak.

It's hard to be sure that models in a wind tunnel really represent what went on in the Devonian era 400 million years ago. It may or may not be true that insect wings began as solar panels and weren't any good for flying until the whole body became bigger for some other reason. It could be that the real physics was different from the models', and the growing stubs were increasingly good for flying right from the start. But Kingsolver and Koehl's research holds a very interesting lesson for us. It teaches us a subtle new way, a kind of sideways diversion, by which paths up Mount Improbable may be found.

For vertebrates the evolution of flight was probably a different story because they are mostly larger anyway. True powered flight has evolved independently in birds, bats (probably separately in at least two different kinds of bats) and pterosaurs. One possibility is that true flight grew out of the habit of gliding between trees, which lots of animals do, even if they don't quite fly. There is a whole world of life in the tree tops. We think of the forest as rising up from the ground. We see it from the vantage point of big, heavy, clumsy, ground-dwelling animals, as we pick our way among the trunks. For us, the deep forest is a cavernous, dark cathedral with arches and vaults stretching up from the ground to a remote green ceiling. But most of the inhabitants of the forest live in the canopy and see the forest from the opposite perspective. Their forest is a vast, gently undulating, sunlit green meadow which, though they hardly notice the fact, just happens to be raised up on stilts. Countless species of animals live their entire lives in this lofty meadow. The meadow is the place where the leaves are, and the leaves are there because that is where the sunlight is, and the sunlight is the ultimate energy source of all life.

The landscape is not literally unbroken. The aerial meadow is pockmarked with holes where it is possible to fall through to the ground: gaps that need to be bridged. In their different ways, many kinds of animals are well equipped to leap across quite large gaps. The difference between a successful leap and an unsuccessful one could be a life and death matter. Any change in body shape that has the effect of extending the leaping range a little further—however little—could well be an advantage. The difference between a squirrel and a rat lies mostly in the tail. The tail is not a wing; you can't fly with it. But it is feathery with hairs that give it a large surface area to catch the air. A rat with a squirrel's tail would undoubtedly be able to leap a larger gap than a rat with a rat's tail. And, if the ancestors of squirrels had rat-like tails, there would be a continuous ramp of improvement, becoming more and more feathery, all the way to a modern squirrel's tail.

I described the squirrel's tail as feathery, but the word is even more appropriate to a totally unrelated little mammal, the feathertail glider (Figure 4.2). This is a marsupial, closer to possums and kangaroos than to rats and squirrels. It lives in the high canopy of the Australian eucalyptus forests. The tail is not, of course, a true feather, which, with its elaborate system of tiny hooks and barbs, is definitively a bird invention. But the feathertail glider's tail looks like a feather and it does a similar job to a feather.

The feathertail glider also has a flap of skin, stretching from elbow to knee, which is capable of extending its leap into a sixty-foot downhill glide. Another family of Australian possums, the flying phalangers, has developed the flap of skin further. In the greater glider, the membrane still only reaches the elbow, but it can nevertheless glide up to 300 feet, and change direction through as much as 90°. The yellow-bellied glider is even more accomplished in the air. Its gliding membrane reaches from wrists to ankles, as does that of the sugar glider, and the larger squirrel glider.

Almost identical in superficial appearance, although utterly unrelated, are the red giant flying squirrel of the Far Eastern forests, and

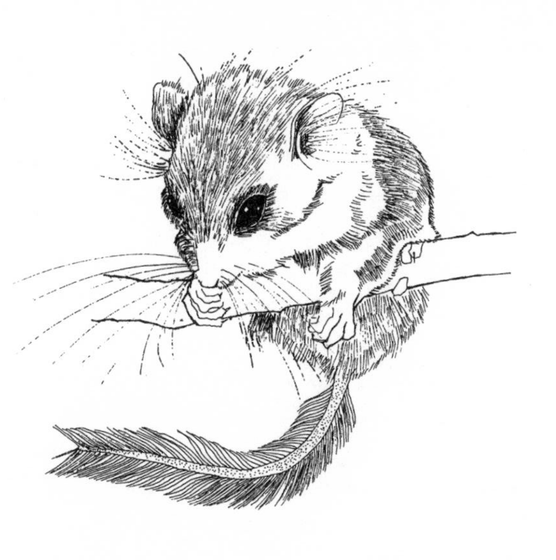

Figure 4.2
Feathertail glider,
Acrobates pygmaeus,
a marsupial from
Australia.

the northern flying squirrel of North America. These are true squirrels—rodents—but, like the more extreme of the marsupial gliders, they have flaps of skin stretching from wrists to ankles. They glide about as capably as their marsupial equivalents. There are other rodents in Africa that have developed the same gliding trick. Although they are called Beecroft's flying squirrel and Zenker's flying squirrel, they are not true squirrels and they have certainly 'invented' gliding independently of the American flying squirrels. An even more comprehensive membrane, which takes in the neck and tail, and the fingers and toes, as well as the arms and legs, is possessed by the mysterious colugo of the Philippines forests. Nobody quite knows what this so-called flying lemur is, except that it isn't a lemur (true lemurs are confined to Madagascar, and none of them fly or glide although several of them leap impressively). Whatever else it may be, the colugo certainly isn't either a rodent or a marsupial. Once again, it has 'invented' the gliding membrane and associated habit entirely independently of all the others.

The colugo, the various flying squirrels and the marsupial gliders all glide with comparable efficiency. But, since the colugo's flight membrane stretches between the fingers, whereas in the others it reaches only as far as the wrists, they could give rise to different kinds of wings if evolution continued further. Even more obviously, the same is true of the beautifully named *Draco volans*, the flying lizard or 'flying dragon' (Figure 4.3). This is a tree-dwelling lizard, also from the forests of the Philippines and Indonesia. Unlike the mammal gliders, its aerial flap does not incorporate the limbs but is stretched between its elongated ribs, which it can erect at will. My favourite of these gliding animals is Wallace's flying frog, a tree frog from the rain forests of South-East Asia. It keeps its flight surface skin between its elongated fingers and toes, and, like the other gliders we have been talking about, it uses it to glide from tree to tree.

In none of these cases is there any difficulty in finding a gentle path up Mount Improbable. Indeed, the fact that the gliding habit has evolved so many times testifies to the ease with which these mountain paths can be found. Perhaps even stronger testimony comes from the paradise tree snake or 'flying snake', again from the South-East Asian forests. This snake makes an effective job of gliding from tree to tree after deliberately throwing itself off, a distance of sixty feet or so, even though it has no obvious sail, flap or flight surface at all. It is just that a snake's drawn-out shape already gives it a relatively large surface area for its weight, and it enhances the effect by pulling in its belly to make a concave surface underneath. This snake would make a perfect first step towards subsequent evolution into something like *Draco volans* with a real gliding membrane. The snake never took the second step, perhaps because elongated ribs would have been an impediment in other aspects of its life.

The way to think of the gradual evolution of something like a flying squirrel is this. To begin with, an ancestor like an ordinary squirrel, living up trees but without any special gliding membrane, leaps across short gaps. However far it can leap without the aid of any special flaps of skin, it could leap a few inches further—and hence save its life when it encounters a gap of critical distance—if it had a very slight flap of skin, or a very slightly increased bushiness of the tail. So natural selection favours individuals with slightly pouchy skin around the arm or leg joints, and this becomes the norm. The normal leaping distance of an average member of the population has thereby

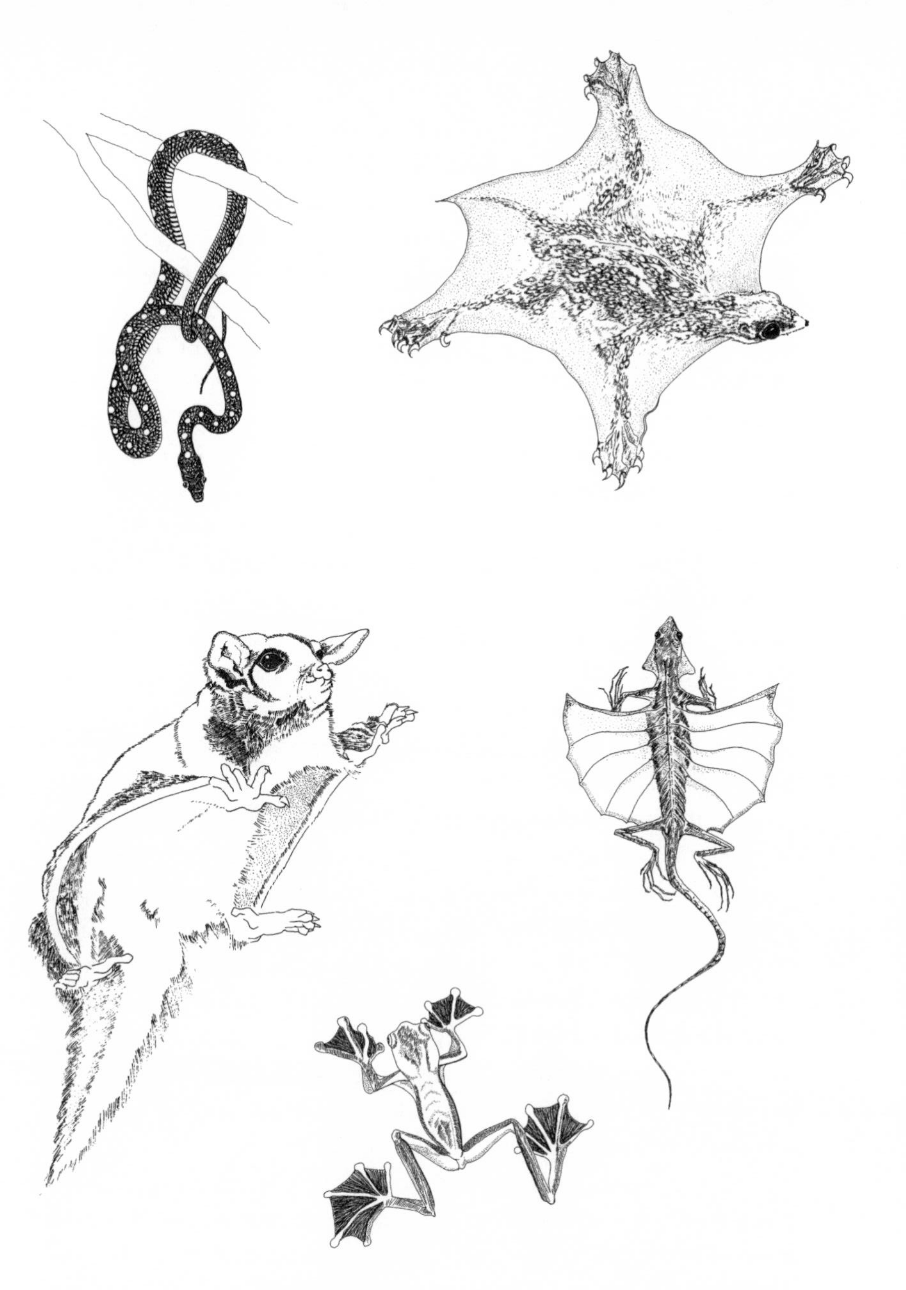

Figure 4.3 Vertebrates that glide down from trees but do not truly fly: (*clockwise from top right*) colugo, *Cynocephalus volans*; flying lizard, *Draco volans*; Wallace's flying frog, *Rhacophorus nigropalmatus*; marsupial sugar glider, *Petaurus breviceps*; and flying snake, *Chrysopelea paradisi*.

been increased by a few inches. Now, any individuals with an even larger skin web can leap a few inches further. So in later generations this extension of skin becomes the norm, and so on. For any given size of membrane, there exists a critical gap such that a marginal increase in the membrane makes all the difference between life and death. The membrane size of the average member of the population gets steadily larger, as the gap that can be jumped by the average member of the population gets larger. After many generations, species like the marsupial gliders and the flying squirrels have evolved, capable of gliding hundreds of feet, and capable of steering themselves into a controlled landing.

But all this gliding still isn't true flying. None of these gliding animals flaps its wings, and none of them can stay in the air indefinitely. They all go downhill, though they may pull up into a short climb by altering their aspect just before landing on a lower tree trunk. It is possible that true flying, as seen in bats, birds and pterosaurs, evolved from gliding ancestors like these. Most of these animals can control the direction and speed of their glide so as to land at a predetermined spot. It is easy to imagine true flapping flight evolving from repetition of the muscular movements used to control glide direction, so average time to landing is gradually postponed over evolutionary time.

Some biologists, however, prefer to see long-distance downhill gliding as the dead end of the tree-jumping line of evolution. True flight, they think, began on the ground rather than up trees. Man-made gliders can take off either by launching off cliffs, or by being pulled rapidly along the ground. Flying fish (Figure 4.4) take off in this second way, though from the sea rather than the land, and they are capable of gliding about the same distance as the best of the marsupial gliders launched from trees. Flying fish swim at great speed in the water and then shoot out into the air, presumably to the consternation of pursuing predators in the water from whose point of view they would vanish, more or less literally, into thin air. They don't hit the water again until they've travelled up to 300 feet. Sometimes when they come down they skim the water with their tails and swim a few strokes to regain speed and take off again. Their 'wings' are greatly enlarged pectoral fins and, in the case of the Atlantic flying fish, enlarged pelvic fins too.

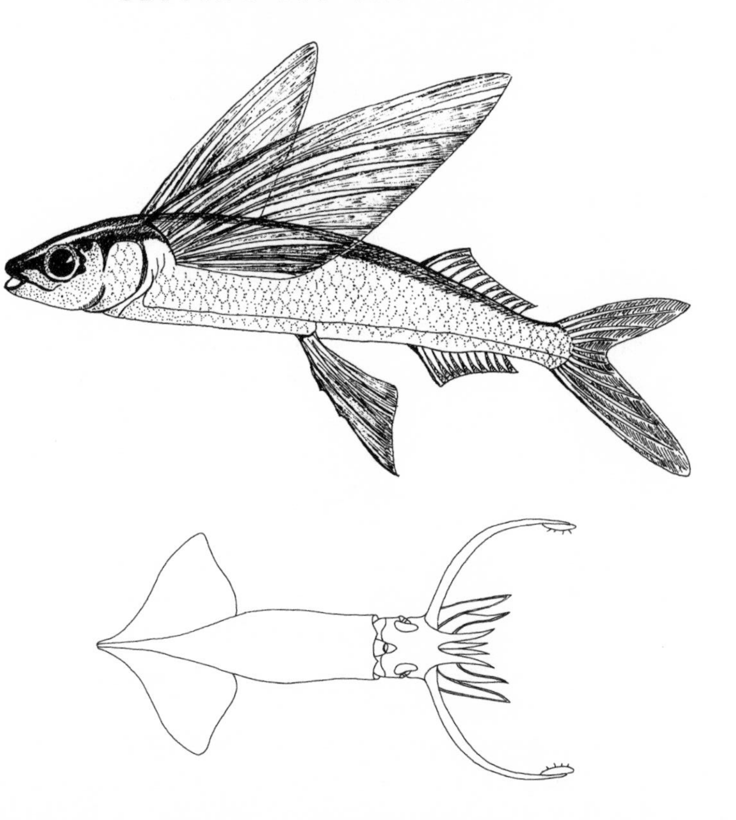

Figure 4.4 Animals that glide after shooting up off the surface. Atlantic flying fish, *Cypselurus heterurus* (top), and flying squid, *Onychoteuthis*.

These true flying fish (Exocoetidae) should not be confused (though they are by at least two books on my coffee table) with the completely unrelated so-called flying gurnards (Dactylopteridae). Far from flying, these gurnards plod along the bottom. They are variously reported to use their 'wings' as stabilizers, as frighteners for flashing at predators, and as stirrers-up of the sand to uncover prey; also, when the fish are disturbed, they rise up into the water a few feet above the bottom, then spread their 'wings' and glide down. The one thing the wings are not used for is flying in air. It is not clear what provoked the legend that they fly: possibly just the large size of their pectoral fins which superficially resemble those of true flying fish. Returning to the true flying fish, they surely evolved not from bottom-dwelling ancestors but from fast-swimming surface fish. Many fish leap out of the water without the aid of enlarged fins. It would surely be potentially easy for such swift leapers to benefit a little by sticking out their fins, and in later generations to increase the area of their fins until they became 'wings'. It is a little sad that dolphins, with their spectacular leaping, have never progressed to the flying-fish stage. Perhaps it is because they'd have to be smaller than existing dolphins in order to do it effectively and there are other reasons, connected with heat insulation and the properties of blubber, why it is hard for warm-blooded dolphins to be small. There are so-called flying squids which behave like flying fish as a means of escaping from the same enemies such as tuna. Squids of the genus *Onychoteuthis* accelerate through the water to speeds of up to forty-five miles per hour, shoot out into the air and glide for more than fifty yards, attaining heights of six feet or more out of the water. They achieve their astonishing speed by jet propulsion, and they fly stern first because, as in all squids, their water jets point past their heads. Once their supply of water has been squirted through the jet, they have no more propulsive force at their disposal until they return to the water. In this respect, flying fish have the advantage because of their habit, already mentioned, of boosting their speed with strokes of the tail while still largely out of the water and skimming the surface.

Fascinatingly, there is one group of fish, the freshwater hatchetfishes of South American rivers, which are reported to vibrate their pectoral fins rapidly and noisily in true, powered flight through the air, although only for short distances. These fishes are not closely related to the true flying fish (nor to the 'flying' gurnards). I must say, I'd like to see a flying hatchet-fish buzzing past my own eyes. I'm not saying I don't believe it; it is agreed by all the books; but, as anglers know and as we learned from the story of the 'flying' gurnards, it is sometimes a good idea to check up on fish stories for oneself.

In any case, I introduced (gliding) flying fish as a prelude to the theory that true, flapping flight evolved not from tree gliders but from fast-running, ground-dwelling animals whose arms became freed from their normal role in running. Flying fish and flying squids, although they live in water, illustrate the principle that if a gliding animal can move sufficiently fast along the surface it can take off without the support of a tree or cliff. The principle might work for birds, because they evolved from two-legged dinosaurs (indeed, you could say that technically birds *are* dinosaurs), some of whom probably ran very fast along the ground, as ostriches do today. To pursue the analogy with flying fish for a moment, the two legs would play the role of the fish's tail, propelling the animal forwards very fast, while the arms play the role of fins, perhaps originally used for stabilizing or steering, and later growing aerofoil surfaces. There are some mammals such as kangaroos that propel themselves very fast on two legs, leaving their arms free to evolve in other directions. Our species seems to be the only mammal to use the two legs in the alternating, bird-like gait, but we are not very fast and we use our arms, not for flying but for carrying things and making things. All the fast-running, twolegged mammals use the kangaroo gait in which the two legs push together rather than alternately. This gait grows naturally out of the horizontal spine-flexing of a typical running quadruped such as a dog. (By analogy, whales and dolphins swim by bending the spine up-and-down, mammal style, whereas fish and crocodiles swim by bending it alternately to left and to right, following the ancient fish habit. Incidentally, we should wonder more than we do at the unsung heroes among the mammal-like reptiles who pioneered the up-and-down gait that we now admire in sprinting cheetahs and greyhounds. Vestiges of the ancient fish wriggle are perhaps still to be seen in tail-wagging dogs, especially when the movement spills over to the whole body in the squirming of a submissive dog.)

Among ground-dwelling mammals, kangaroos and their marsupial kind don't have a monopoly of the 'kangaroo gait'. My colleague Dr Stephen Cobb was once lecturing to zoology students in the University of Nairobi and he told them that wallabies are confined to Australia and New Guinea. 'No, sir,' a student protested. 'I have seen one in Kenya.' What the student had seen was undoubtedly one of these (Figure 4.5).

This animal, the so-called springhaas or spring hare, is neither a hare nor a kangaroo but a rodent. Like kangaroos, it hops to increase its speed when fleeing from predators. Other rodents like the jumping jerboa do the same. But bipedal mammals don't seem to have taken the next step and evolved the power of flight. The only true flying mammals are bats, and their wing membrane incorporates the back legs as well as the arms. It is hard to see how such a leg-encumbering wing could have evolved by the fast-running route. The same is true of pterosaurs. My guess is that both bats and pterosaurs evolved flight by gliding downwards from trees or cliffs. Their ancestors, at one stage, might have looked a little like colugos.

Birds could be another matter. Their story is different anyway, centred around that wondrous device, the feather. Feathers are modified reptilian scales. It is possible that they originally evolved for a different purpose for which they are still very important, heat insula-

Figure 4.5 Spring hare, *Pedetes capensis.*

tion. At all events, they are made of a horny material which is capable of forming light, flat, flexible yet stiff flight surfaces. Bird wings are very unlike the saggy skin flaps of bats and pterosaurs. The ancestors of birds therefore were capable of forming a proper wing which didn't have to be stretched between bones. It was enough to have a bony arm at the front. The stiffness of the feathers themselves took care of the rest. The back legs could be left free to run. Far from being awkward and clumsy on the ground like bats and, presumably, pterosaurs, birds can use their legs for running, jumping, perching, climbing, prey-catching and fighting. Parrots even use their feet like human hands. Meanwhile the front limbs get on with the business of flying.

Here's one guess as to how flying got started in birds. The hypothetical ancestor, which we can imagine as a small, agile dinosaur, runs fast after insects, leaping in the air with its powerful hind legs and snapping at the prey. Insects had evolved into the air long before. A flying insect is perfectly capable of taking evasive action, and the leaping predator would benefit from skill in mid-course correction. To some extent you can see cats doing this today. It seems difficult because, since you are in the air, there is nothing solid to push against. The trick is to shift your centre of gravity. You can do this by moving bits of yourself relative to other bits. You could move your head or tail, but the obvious bits to move are the arms. Now, once the arms are being moved for this purpose, they become more effective at it if they develop surfaces to catch the air. It has also been suggested that the feathers on the arms originally developed as a kind of net for catching insects. This is not so far-fetched as it sounds, for some bats use their wings in this way. But, according to this theory, the most important use of the arms was for steerage and control. Some calculations suggest that the most appropriate arm movements for controlling pitching and rolling in a leap would actually resemble rudimentary flapping movements.

The running, jumping and mid-course correction theory, when compared with the tree-gliding theory, reverses the order of things. On the tree-gliding theory, the original role of the proto-wings was to provide lift. Only later were they used for control, and then finally flapping. On the jumping for insects theory, control came first, and only later were the arms with their surfaces commandeered to provide lift. The beauty of this theory is that the same nervous circuits as were used to control the centre of gravity in the jumping ancestor would, rather effortlessly, have lent themselves to controlling the flight surfaces later in the evolutionary story. Perhaps birds began flying by leaping off the ground, while bats began by gliding out of trees. Or perhaps birds too began by gliding out of trees. The debate continues.

In any case modern birds have come a long way since those early days. Lots of long ways I should say, for the peaks of Mount Improbable that they have conquered are splendidly many. A peregrine falcon can dive out of the sky at more than 100 miles per hour when closing in on prey. Hawks and humming-birds hover with pinpoint precision in one spot like a helicopter's wildest dream. Arctic terns spend more than half of every year on their annual migration from Arctic to Antarctic and back again, a distance of 12,000 miles. The wandering albatross, slung below its ten-foot wing-span, circles the pole with an ever clockwise heading, powered not by flapping but by vigilant attention to the natural engine of changing wind speed where the cold waves shear the Roaring Forties. Some birds, like pheasants and peacocks, use flight only in an occasional explosive burst when startled by possible danger. Others, like ostriches, rheas and the lamented giant moas of New Zealand, became too big to fly and their wings degenerated in comparison to their huge, striding and kicking legs. At the other extreme, swifts have feeble, clumsy legs but state-of-the-art swept-back wings and they almost never leave the air. They land only to nest, even mating and sleeping on the wing. When they do land they must choose a high place, for they cannot take off from level ground. They build their nests from materials that they meet floating in the air, or snatch from trees while screaming past. For a swift, coming to earth may seem like a difficult, unnatural state, comparable to, say, sky-diving for a human, or swimming underwater. For us, the world is a steady, motionless backdrop to our preoccupations. Through a swift's black eyes, the normal, background state of the world is a ceaselessly hurtling horizon, dizzily tilting by. Our terra firma may be the swift's idea of an unnerving Disneyland roller-coaster.

Not all birds flap their wings, but those that soar or glide have probably come from flapping ancestors. Flapping flight is complicated and not, in every detail, understood. It is tempting to think that the powerful downward beats of the wings provide lift directly. This may be a part of the story, especially during take-off, but most of the lift is provided by the shape of the wings (given enough air speed), as in an aeroplane. A specially curved or tilted wing can provide lift if a wind is blown over it or—which amounts to the same thing—if the whole bird is moving forwards relative to the air, for whatever reason. The flapping movement of the wings is mainly concerned with providing the necessary forward thrust. This propeller role of the wings relies upon the fact that they don't simply flap up and down. Instead, the bird imparts an artful twist from the shoulder together with subtle adjustments at all the joints, and some other benefits follow automatically from the bending of the feathers. As a result of these twists, adjustments and bendings, the up-and-down stroking movements of the wings are translated into forward thrust, slightly in the manner of a whale's up-and-down beating tail. Given that there is forward movement through the air, the wings of a bird provide lift in approximately the same way as those of an aeroplane, albeit aeroplane wings are simpler because fixed. The higher the speed, the greater the lift, which is why a Boeing 747 stays airborne in spite of its Brobdingnagian weight.

The laws of physics conspire to make flapping flight increasingly difficult for larger birds. If we think of birds of the same shape getting progressively larger, weight goes up as the cube of length but wing area only goes up as the square of length. In order to stay in the air, larger birds would need to grow disproportionately larger wings, and/or fly disproportionately faster. As we imagine ever larger and larger birds, there comes a point where, lacking jet or piston engines, the muscle power available to a bird of that size is no longer strong enough to keep it in the air. This critical point in the size range is somewhat smaller than the larger vultures and albatrosses. Some large birds, as we have seen, simply gave up the struggle, grounded themselves for good, and made a nice living getting even larger as ostriches and emus. But vultures, condors, eagles and albatrosses are not grounded. Why not?

Their trick is to exploit external energy sources. If it weren't for the sun's heat and the moon's shifting gravity, air and sea would be still. External energy charges up the ocean currents, pumps up the winds, stirs the dust-devils, rocks the atmosphere with powerful forces capable of flattening a house or driving a trade-route; it also engenders thermal up-currents which, if you use them judiciously, can raise you to the clouds. Vultures, eagles and albatrosses use them to perfection. They may be the only animals to match our skill in mining the energy of the weather. My main source of information on soaring birds is the writings of Dr Colin Pennycuick of Bristol University. He used his specialized knowledge as a glider pilot, both to understand how birds do it and to glide among them in order to study their techniques in the field.

Vultures and eagles use thermals, just as human glider pilots do. A thermal is a column of warm air that rises, perhaps because a patch of ground at its base soaks up more than its share of the sunlight. Glider pilots are largely dependent on thermals, and experience makes them expert in spotting them from a distance. The subtle cues that betray a thermal's presence include certain characteristic shapes of cumulus cloud at their tops, and certain conformations of ground at their base. The approved technique for cross-country gliding is to circle your way up to the top of a thermal, say a mile high, then launch out in a straight downward glide in the direction you want to travel. The ramp down is a shallow one: a vulture typically loses one yard of height for every ten yards of forward travel. This gives it nearly ten miles of cross-country travel before it needs to find another thermal and rack itself up to the top again.

As it happens, thermals are often arranged in 'streets', which a glider pilot can see ahead by reading the clouds. Vultures too, like human glider pilots, are adept at following these streets. Sometimes, when a vulture finds a good street lined up in the direction it wants to go, it glides along the street gaining lift from each thermal without bothering to circle in it. In this way, a vulture can travel great distances without pausing to circle. They usually do this only when commuting from feeding grounds to nesting site. Most of the time vultures are not travelling long distances in straight lines but cruising around in search of carrion. They also keep an eye on other vultures. If any one finds a carcass and goes down, others notice and quickly join it. In this way, a wave of attention spreads across the sky, like the wave of fires lit on hilltops that spread the length and breadth of England to warn of the Spanish Armada.

A similar trick of watching companions is used by white storks for a different purpose in their long annual migration from northern Europe to southern Africa. They travel in bands of up to some hundreds. Like vultures, they wind themselves up to the tops of thermals, then head off across country until they find another thermal. But although they circle together in a thermal, when they leave it, instead of travelling in a close-formation flock they spread out in line abreast. With such a broad front advance, if they just glide straight some of them are very likely to find another thermal. When they do, their neighbours in the line notice them rising, and come to join them. In this way, all the members of the extended group benefit from thermals that any members find.

Whichever view we take of the origin of flight in birds, whether the tree-gliding theory or the running and leaping theory, vultures and eagles, storks and albatrosses, are almost certainly secondary gliders. They evolved their gliding technique from ancestors that flapped and were smaller. For the school of thought that sees bird flight as originating in gliding from trees, modern vultures—albeit they soar up thermals rather than climb trees to gain height—would represent a *return* to gliding via an intermediate stage of flapping. During this intermediate flapping stage their nervous systems would, on this theory, have acquired new circuitry and new skills of control and manoeuvre. These new skills would have left them with improved efficiency when they returned to non-flapping flight. It is quite common for animals to return to a much older way of life, having served an evolutionary apprenticeship in another one, and it may be plausible to argue that they return from the apprenticeship better equipped to tackle the original way of life. Soaring birds may not be a good example, because it is uncertain how bird flight originally began. A more clear-cut example of animals returning to an earlier way of life is provided by those that have returned to the water, having spent some millions of years on land. It is to these that I now turn, as a coda to this chapter (Figure 4.6).

Fifty million years ago, the ancestors of whales and sea-cows (dugongs and manatees) were land-dwelling mammals, probably carnivores in the case of whales, herbivores in the case of sea-cows. The ancestors of these and all other land-dwelling mammals had been, much earlier still, sea-dwelling fish. The return to the water by the whales and dugongs was a homecoming. As always, we can be sure that it took place gradually. They took to the water, perhaps at first just to feed, like a modern otter. They must have spent progressively less and less time on the land, perhaps going through a phase where they resembled modern seals. Now they never leave the water and are completely helpless if beached. Nevertheless, they bear numerous reminders of their land-dwelling ancestors and they also, like all mammals, have much older relics of their previous incarnation in the water. Whales breathe air, for their landlubber ancestors lost the use of their gills. But all mammals, whales and sea-cows included, have traces of gills in the embryo: unmistakable vestiges of their remote past in the water. Freshwater snails, too, have gone back to water from the land, and they breathe air. Their earlier ancestors lived in the sea, like most of the snail family today. Snails seem to have gone from sea to fresh water via a land 'bridge': perhaps something about land life eases the transition. Other land animals that have gone back to the water include turtles, water beetles, diving-bell spiders, and the extinct ichthyosaurs and plesiosaurs. Turtles do manage to extract some oxygen from the water, but they do it not with gills but with the lining of the mouth, and in some cases the lining of the rectum and, in soft-shelled turtles, the skin covering the shell. Water beetles and spiders take a bubble of air down with them. All these animals are returning to the watery environment of their more remote ancestors, but when they get there they do things differently because of their interim race experience.

When land animals return to the water, why don't they rediscover the full apparatus of watery living? Why don't whales and sea-cows

Figure 4.6 Whales and sea-cows. Animals that have returned to the sea after hundreds of millions of years on land: (from top) dugong, *Dugong dugon*; manatee, *Trichechus senegalensis*; humpback whale, *Megaptera novaeangliae*; killer whale, *Orcinus orca*.

regrow gills and lose their lungs? This brings us to another important lesson that Mount Improbable has to teach us. In evolution, ideal outcomes are not the only consideration. It also makes a difference where you start: as in the story of the man who, when asked the way to Dublin, replied, 'Well, I wouldn't start from here.' Mount Improbable has many peaks. There are many ways of living in water. You can use gills to get oxygen from the water, or you can come to the surface and breathe air. Continually coming to the surface seems like an oddly inconvenient habit. Maybe it is, but the ancestors of whales and sea-cows began, remember, close to an air-breathing peak of the mountain. All their internal details were geared to air-breathing assumptions. Perhaps they could have reformed them and come into line with the fishes, dusting off the embryonic vestiges of their ancient gills. But that would have meant a massive shake-up of their bodily infrastructure. It would have been equivalent to going down a deep valley between two peaks of the mountain, with the ultimate objective of climbing a slightly higher peak eventually. It cannot be said too often that Darwinian theory does not allow for getting temporarily worse in quest of a long-term goal.

Even if they had gone down the valley, it isn't necessarily obvious that the gill peak, when they eventually climbed it, would have turned out to be higher. Gills are not necessarily better than lungs for water- dwelling animals. No doubt it is convenient to be able to breathe continuously, wherever you are, rather than having to break off what you are doing to go to the surface. But our judgement is coloured by the fact that we take a breath every few seconds and panic at even a brief interruption to our air supply. Having been naturally selected through millions of sea-going generations, sperm whales can submerge for fifty minutes before they have to breathe. Coming to the surface to breathe, for a whale, might feel rather like going off to urinate. Or for a meal. If you start to think of breaths as meals, rather than as a continuously vital necessity, it becomes less obvious that every underwater creature would ideally be better off with gills. There are animals, like humming-birds, that feed more or less continuously. To a humming-bird, which needs to suck nectar every few seconds of its waking life, visiting flowers might feel rather like breathing. Sea-squirts, bag-shaped marine invertebrates remotely related to vertebrates, pump a never-ceasing current of water through their bodies, filtering out tiny particles of food. Such a filter-feeder indulges in nothing corresponding to a meal. A sea-squirt might suffocate with panic at the thought of having to search for the next meal. Sea-squirts might well wonder why so many animals go in for the absurdly inefficient and dangerous habit of searching for meals, instead of sitting back and breathing in food the whole time.

Be this as it may, there is no doubt that whales and dugongs come with their dry-land history written all over them. If they had been deliberately created for the sea, they'd be very different, and a lot more like fish than they are. Animals that have their history written all over them are among the most graphic pieces of evidence we have that living things were not created for their present ways of life but evolved from very different ancestors.

Plaice, sole and flounders have their history written all over them too, to the point of grotesqueness. No sane creator, setting out from scratch to design a flat-fish, would have conceived on his drawing board the absurd distortion of the head needed to bring both eyes round to one side. He'd surely, right from the start, have gone for the skate or ray design, the fish lying on the belly with the eyes symmetrically placed on the top (Figure 4.7). Plaice and sole are all twisted around because of their history; because their ancestors lay down on one side. Skates and rays are gracefully symmetrical because their history happened to be different: when their ancestors settled for sea-floor life they lay on the belly rather than on their side. When I say 'happened to be' different, I don't mean that there wasn't a good reason for the difference. Skates and rays are descended from sharks, and sharks are already slightly flattened compared with bony fish which are typically deep-bodied and blade-like. A deep-bodied blade of a fish can't lie on its belly, it has to flop over on its side. When they settled on the bottom, the ancestors of plaice raced up the nearest peak of Mount Improbable, regardless of the fact that a possibly

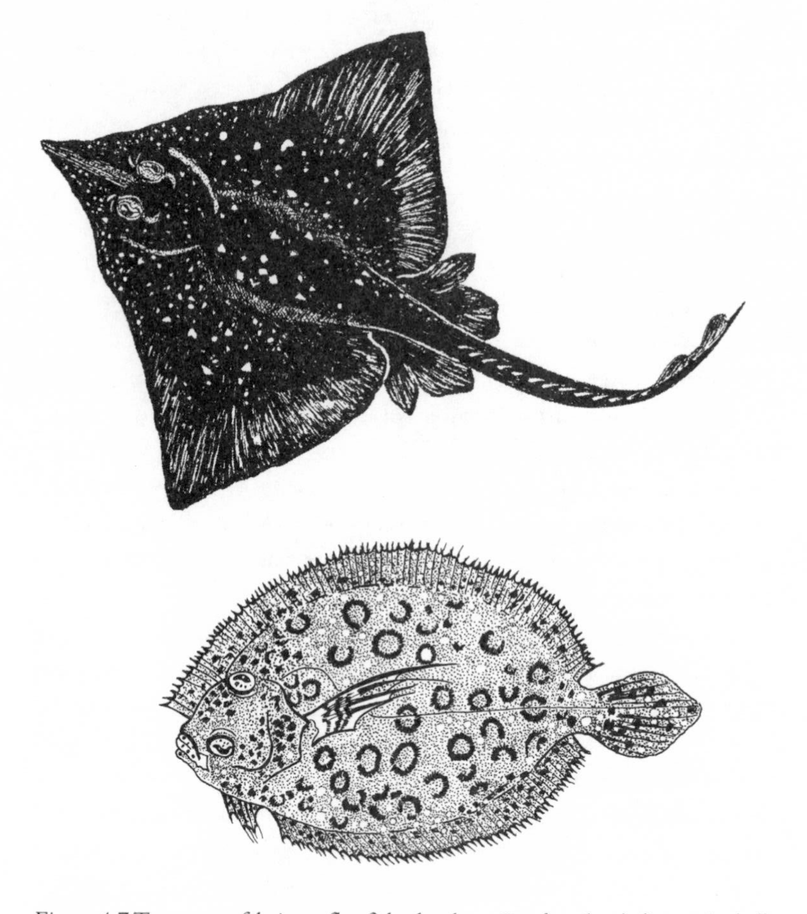

Figure 4.7 Two ways of being a flat-fish: the skate, *Raja batis* (top), lies on its belly, while the flounder, *Bothus lunatus*, lies on its side.

higher peak of the mountain—the skate/ray symmetrical peak—was theirs for the taking, if only they could force their way down a little valley in order to reach the foot of the higher peak. To say it again, going down the slopes of Mount Improbable is not allowed by natural selection, and these fish had no choice but to restore their vision in a makeshift way by twisting one eye round to the other side of the body. The ancestors of skates also rushed up their nearest flat-fish peak, which in their case led them to their symmetrical elegance. Of course when I speak of having no 'choice', and 'rushing' up mountain peaks, you understand, as usual, that it is not individual fish that are meant. It is evolutionary lineages, and 'choice' refers to available alternative routes of evolutionary change.

I've stressed that going downhill is not allowed, but not allowed by whom? And can it *never* happen? The answer to both questions is about the same as for the case of a river not being 'allowed' to run in any direction other than along its established watercourse. Nobody actually orders the water to stay within the banks of a river but, for well-understood reasons, it normally does. Just occasionally, however, it overflows the banks, or even bursts them, and the river may be found to have altered its permanent course as a result.

What might permit an evolving lineage to go into reverse for a brief while and so expose itself to the opportunity to ascend a previously inaccessible peak of Mount Improbable? This is the kind of question that interested the great geneticist Sewall Wright who, by the way, was the first to use a landscape metaphor for evolution, the progenitor of my Mount Improbable. Wright was the American member of the cantankerously warring triumvirate who, in the 1920s and 1930s, founded what we now call neo-Darwinism. (The other two were English, those incomparable but bellicose prodigies R. A. Fisher and J. B. S. Haldane, and it is only fair to add that the cantankerousness seems all to have originated from them, not from Wright.) Wright realized that, paradoxically, natural selection can be a force *against* extreme perfection. This is for precisely the reason that we have just been dealing with. Going down valleys is forbidden by natural selection. A species that is trapped on a small foothill of the mountain cannot escape to higher peaks as long as natural selection pens it in at the top of the foothill. If only natural selection would relax its grip for a short while, the species might edge its way down the foothill far enough to cross a valley to the lower slopes of a high peak. Once there, it is in a position, when natural selection starts biting again, to evolve rapidly up the higher slopes of the mountain. From a global point of view, then, one recipe for improvement in evolution is periods of strong selection interrupted by periods of relaxation. Maybe this kind of relaxation is actually important in real-life evolution.

When might 'relaxation' occur? One possibility is when there is a vacuum to be filled. In a small way this will happen whenever a population is growing because it is smaller than the area can support. There may be bonanzas of opportunity and relaxation of selection when a virgin continent is first colonized after being cleaned out by a catastrophe. Perhaps after the dinosaurs became extinct, the remaining mammals had such a field day of opportunity that some of their lineages 'relaxed their guard', went temporarily downhill, and thereby found higher peaks of Mount Improbable from which they would normally have been debarred.

Another recipe is transfusions of fresh genes from other places. This is the point that I said I'd return to from Chapter 2 on spider webs. In the NetSpinner model of spider webs, there was not just one sexual population of simulated web-spinners but three 'demes' evolving in parallel. These were thought of as evolving independently in three different geographical areas. But—here's the point—not *completely* independently. There is a trickle of genes, meaning that an individual occasionally migrates, from one local population to another. The way I put it was that these migrant genes were a kind of injection of fresh 'ideas' from another population: 'almost as though a successful sub-population sends out genes that "suggest" to a less successful population a better way to solve the problem of building a web'. It is tantamount to being guided up the higher peak of the metaphorical mountain by a smuggled-in map.

We are ready to take up the creationists' favourite target, and the star stumbling block for would-be believers in evolution, perched precariously on the summit of the most formidable cliff Mount Improbable has to offer: the eye.

Note: After this book had gone to be typeset, J. H. Marden and M. G. Kramer published a fascinating study of stoneflies, which suggests yet another possible route up Mount Improbable towards true flapping flight (Marden, J. H., & Kramer, M. G. (1995) 'Locomotor performance of insects with rudimentary wings'. *Nature,* 377, 332–4). Stoneflies are rather primitive flying insects, where primitive means that, although they are modern living insects, they are thought to resemble ancestors more than other modern insects resemble ancestors. The particular species that Marden and Kramer studied, *Allocapnia vivipara,* skims across the surface of streams by raising its wings and using them as sails to catch the wind. Sailing velocity is approximately proportional to wing length. Individuals with the smallest wings sail faster than individuals who don't raise their wings at all. Those same smallest wings are roughly the same size as the movable gill plates of early fossil insects. Maybe wingless ancestors lived on water surfaces and raised their gill plates as sails. There would then have been a smooth ramp up Mount Improbable as the gill plates grew to become progressively more effective sails. As for the next step towards flapping flight on this hypothesis, Marden and Kramer have made another relevant observation. A different species of stonefly, *Taeniopteryx burksi,* also skims along the water surface, but it flaps its wings to do so. Perhaps insects, on their way up to the flying peak of Mount Improbable, passed through a sailing phase like *Allocapnia,* then a surface flapping phase like *Taeniopteryx.* It is easy to imagine that light flapping insects buzzing their way over the surface might occasionally have been lifted by gusts of wind. There would then have been a further ramp up the mountain as their flapping wings kept them aloft for progressively longer times.

CHAPTER 5

THE FORTY-FOLD PATH TO ENLIGHTENMENT

ALL ANIMALS HAVE TO DEAL WITH THEIR WORLD, AND the objects in it. They walk on objects, crawl under them, avoid crashing into them, pick them up, eat them, mate with them, run away from them. Back in the geological dawn when evolution was young, animals had to make physical contact with objects before they could tell that those objects were there. What a bonanza of benefit was waiting for the first animal to develop a remote-sensing technology: awareness of an obstacle before hitting it; of a predator before being seized; of food that wasn't already within reach but could be anywhere in the large vicinity. What might this high technology be?

The sun provided not only the energy to drive the chemical cogwheels of life. It also offered the chance of a remote guidance technology. It pummelled every square millimetre of Earth's surface with a fusillade of photons: tiny particles travelling in straight lines at the greatest speed the universe allows, criss-crossing and ricocheting through holes and cracks so that no nook escaped, every cranny was sought out. Because photons travel in straight lines and so fast, because they are absorbed by some materials more than others and reflected by some materials more than others, and because they have always been so numerous and so all-pervading, photons provided the opportunity for remote-sensing technologies of enormous accuracy and power. It was necessary only to detect photons and—more difficult—distinguish the directions from which they came. Would the opportunity be taken up? Three billion years later you know the answer, for you can see these words.

Darwin famously used the eye to introduce his discussion on 'Organs of extreme perfection and complication':

> To suppose that the eye, with all its inimitable contrivances for adjusting the focus to different distances, for admitting different amounts of light, and for the correction of spherical and chromatic aberration, could have been formed by natural selection, seems, I freely confess, absurd in the highest possible degree.

It is possible that Darwin was influenced by his wife Emma's difficulties with this very point. Fifteen years before *The Origin of Species* he had written a long essay outlining his theory of evolution by natural selection. He wanted Emma to publish it in the event of his death and he let her read it. Her marginalia survive and it is particularly interesting that she picked out his suggestion that the human eye 'may possibly have been acquired by gradual selection of slight but in each case useful deviations'. Emma's note here reads, 'A great assumption / E.D.' Long after *The Origin of Species* was published Darwin confessed, in a letter to an American colleague: 'The eye, to this day, gives me a cold shudder, but when I think of the fine known gradations, my reason tells me I ought to conquer the cold shudder.' Darwin's occasional doubts were presumably similar to those of the physicist whom I quoted at the beginning of Chapter 3. Darwin, however, saw his doubts as a challenge to go on thinking, not a welcome excuse to give up.

When we speak of 'the' eye, by the way, we are not doing justice to the problem. It has been authoritatively estimated that eyes have evolved no fewer than forty times, and probably more than sixty times, independently in various parts of the animal kingdom. In some cases these eyes use radically different principles. Nine distinct principles have been recognized among the forty to sixty independently evolved eyes. I'll mention some of the nine basic eye types—which we can think of as nine distinct peaks in different parts of Mount Improbable's *massif*—as I go on.

How, by the way, do we ever know that something has evolved independently in two different groups of animals? For example, how do we know that bats and birds developed wings independently? Bats are unique among mammals in having true wings. It could theoretically be that ancestral mammals had wings, and all except bats have subsequently lost them. But for that to occur, an unrealistically large number of independent wing losses would be required, and the evidence supports common sense in indicating that this didn't happen. Ancestral mammals used their front limbs not for flying but for walking, as the majority of their descendants still do. It is by means of similar reasoning that people have worked out that eyes have arisen many times independently in the animal kingdom. We can also use other information such as details of how the eyes develop in the embryo. Frogs and squids, for instance, both have good camera-style eyes, but these eyes develop in such different ways in the two embryos that we can be sure they evolved independently. This does not mean that the common ancestor of frogs and squids totally lacked eyes of any kind. I wouldn't be surprised if the common ancestor of all surviving animals, who lived perhaps a billion years ago, possessed eyes. Perhaps it had some sort of rudimentary patch of light-sensitive pigment and could just tell the difference between night and day. But eyes, in the sense of sophisticated image-forming equipment, have evolved many times independently, sometimes converging on similar designs, sometimes coming up with radically different designs. Very recently there has been some exciting new evidence bearing upon this question of the independence of the evolution of eyes in different parts of the animal kingdom. I'll return to it at the end of the chapter.

As I survey the diversity of animal eyes, I'll often mention whereabouts on the slopes of Mount Improbable each type is to be found. Remember, though, that these are all eyes of modern animals, not of true ancestors. It is convenient to think that they might give us some clues about the kinds of eyes that ancestors had. They do at least show that eyes that we think of as lying half-way up Mount Improbable would actually have worked. This really matters for, as I have already remarked, no animal ever made a living by being an intermediate stage on some evolutionary pathway. What we may think of as a way station up the slope towards a more advanced eye may be, for the animal itself, its most vital organ and very probably the ideal eye for its own particular way of life. High-resolution image-forming eyes, for instance, are not suitable for very small animals. High-quality eyes have to exceed a certain size—absolute size not size relative to the animal's body—and the larger the better in absolute terms. For a very small animal an absolutely large eye would probably be too costly to make and too heavy and bulky to carry around. A snail would look pretty silly if its eyes had the seeing power of human eyes (Figure 5.1). Snails that grew eyes even slightly larger than the present average might see better than their rivals. But they'd pay the penalty of having to carry a larger burden around, and therefore wouldn't survive so well. The largest eye ever recorded, by the way, is a colossal 37 cm in diameter. The leviathan that could afford to carry such eyes around is a giant squid with 10-metre tentacles.

Accepting the limitations of the metaphor of Mount Improbable, let's go right down to the bottom of the vision slopes. Here we find

Figure 5.1 Fantasy snail with eyes large enough to see as well as humans can.

eyes so simple that they scarcely deserve to be recognized as eyes at all. It is better to say that the general body surface is slightly sensitive to light. This is true of some single-celled organisms, some jellyfish, starfish, leeches and various other kinds of worms. Such animals are incapable of forming an image, or even of telling the direction from which light comes. All that they can sense (dimly) is the presence of (bright) light, somewhere in the vicinity. Weirdly, there is good evidence of cells that respond to light in the genitals of both male and female butterflies. These are not image-forming eyes but they can tell the difference between light and dark and they may represent the kind of starting point that we are talking about when we speak of the remote evolutionary origins of eyes. Nobody seems to know how the butterflies use them, not even William Eberhard, whose diverting book, *Sexual Selection and Animal Genitalia,* is my source for this information.

If we think of the plain below Mount Improbable as peopled by ancestral animals that were totally unaffected by light, the non-directional light-sensitive skins of starfish and leeches (and butterfly genitals) are just a little way up the lower slopes, where the mountain path begins. It is not difficult to find the path. Indeed it may be that the 'plain' of total insensitivity to light has always been small. It may be that living cells are more or less bound to be somewhat affected by light—a possibility that makes the butterfly's light-sensitive genitals seem less strange. A light ray consists of a straight stream of photons. When a photon hits a molecule of some coloured substance it may be stopped in its tracks and the molecule changed into a different form of the same molecule. When this happens some energy is released. In green plants and green bacteria, this energy is used to build food molecules, in the set of processes called photosynthesis. In animals the energy may trigger a reaction in a nerve, and this constitutes the first step in the process called seeing, even in animals lacking eyes that we would recognize as eyes. Any of a wide variety of coloured pigments will do, in a rudimentary way. Such pigments abound, for all sorts of purposes other than trapping light. The first faltering steps up the slopes of Mount Improbable would have consisted in the gradual improvement of pigment molecules. There is a shallow, continuous ramp of improvement—easy to climb in small steps.

This lowland ramp pushed on up towards the evolution of the living equivalent of the photocell, a cell specialized for capturing photons with a pigment, and translating their impact into nerve impulses. I shall continue to use the word photocell for those cells in the retina (in ourselves they are called rods and cones) which are specialized for capturing photons. The trick that they all use is to increase the number of layers of pigment available to capture photons. This is important because a photon is very likely to pass straight through any one layer of pigment and come out the other side unscathed. The more layers of pigment you have, the greater the chance of catching any one photon. Why should it matter how many photons are trapped and how many get through? Aren't there always plenty of photons to spare? No, and the point is fundamental to our understanding of the design of eyes. There is a kind of economics of photons, an economics as mean-spirited as human monetary economics and involving inescapable trade-offs.

Before we even get into the interesting economic trade-offs, there can be no doubt that in absolute terms photons are in short supply at some times. On a crisp, starry night in 1986 I woke my two-year-old daughter Juliet and carried her, wrapped in blankets, out into the garden where I pointed her sleepy face towards the published location of Halley's Comet. She didn't take in what I was saying, but I stubbornly whispered into her ear the story of the comet and the certainty that I could never see it again but that she might when she was seventy-eight. I explained that I had woken her so that she'd be able to tell her grandchildren in 2062 that she had seen the comet before, and perhaps she'd remember her father for his quixotic whim in carrying her out into the night to show it to her. (I may even have whispered the words quixotic and whim because small children like the sound of words they don't know, carefully articulated.)

Probably some photons from Halley's Comet did indeed touch Juliet's retinas that night in 1986 but, to be truthful, I had a hard time convincing myself that I could see the comet. Sometimes I seemed to conjure a faint, greyish smear at approximately the right place. At other times it melted away. The problem was that the number of photons falling on our retinas was close to zero.

Photons arrive at random times, like raindrops. When it is really raining properly we are in no doubt of the fact and wish our umbrella hadn't been stolen. But when rain starts gradually, how do we decide the exact moment when it begins? We feel a single drop and look up apprehensively, unconvinced until a second or a third drop arrives. When rain is spitting infrequently like this, one person may say that it is raining while his companion denies it. The drops can fall infrequently enough to hit one person a minute before his companion registers a hit. To be really convinced that there is light, we need the photons to patter on our retinas at an appreciably high rate. When Juliet and I gazed in the general direction of Halley's Comet, photons from the comet were probably hitting individual photocells on our retinas at the fantastically slow rate of about one every forty minutes! This means that any one photocell could be saying, 'Yes there is light there,' while the vast majority of its neighbouring photocells were not. The only reason I received any sensation at all of a comet-shaped object was that my brain was summing up the verdicts of hundreds of photocells. Two photocells capture more photons than one. Three capture more than two, and so on up the slope of Mount Improbable. Advanced eyes like ours have millions of photocells densely packed like pile in a carpet, and each one of them is set up to capture as many photons as possible.

Figure 5.2 is a typical advanced photocell, from a human as it happens, but others are much the same. The writhing colony of apparent maggots in the middle of the picture are mitochondria. These are small bodies that live inside cells. They are originally descended from parasitic bacteria but they have made themselves indispensable for producing energy in all our cells. The nervous connecting wire of the photocell leads off at the left of the picture. The elegant rectangular array of membranes, lined up with military precision on the right, is where the photons are trapped. Each layer contains molecules of the vital, photon-trapping pigment. I count ninety-one layers of membrane in this picture. The exact number is not critical: the more the merrier as far as catching photons is concerned, though there will be overhead costs prohibiting too many layers. The point is that ninety-one membranes are more effective in stopping photons than ninety, ninety are more effective than eighty-nine, and so on back to one

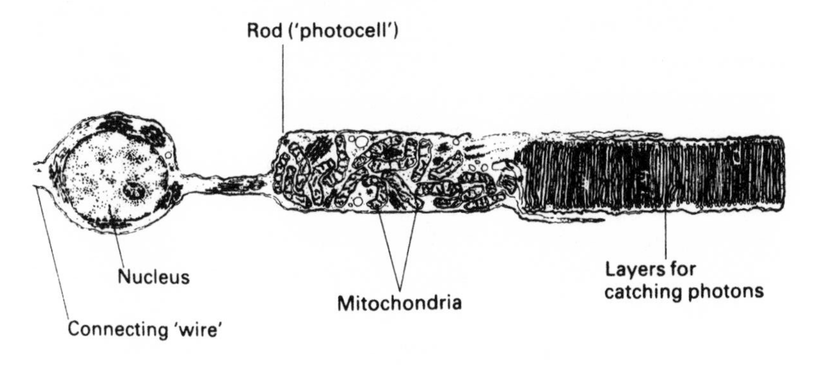

Figure 5.2 Photon-capturing device or 'biological photocell': a single retinal cell (rod) of a human.

membrane, which is more effective than zero. This is the kind of thing I mean when I say that there is a smooth gradient up Mount Improbable. We would be dealing with an abrupt precipice if, say, any number of membranes above forty-five was very effective while any number below forty-five was totally ineffective. Neither common sense nor the evidence leads us to suspect any such sudden discontinuity.

Squids, as we have seen, evolved their similar eyes independently of vertebrates. Even their photocells are very alike. The main difference is that in the squid the layers, instead of being packed as a stack of discs, are rings stacked around a hollow tube. (This kind of superficial difference is common in evolution, for the same kind of inconsequential reason as, say, the fact that English light switches go down for on, American ones down for off.) All advanced animal photocells are playing different versions of the same trick of increasing the number of layers of pigment-laced membranes through which a photon must pass if it would escape untrapped. From Mount Improbable's point of view, the important point is that one more layer marginally improves the chance of trapping photons no matter how many, or how few, layers there already are. Ultimately, when most of the photons have been caught, there will be a law of diminishing returns for the increasing cost of more layers.

Of course there is not much call, in the wild, for detecting Halley's Comet, returning every seventy-six years with its negligible contribution of reflected photons. But it is very useful to have eyes sensitive enough to see by moonlight and even starlight if you are an owl. On a typical night, any one of our photocells might receive photons at about one per second, admittedly a higher rate than for the comet, but still slow enough to make it vital to trap every last photon if it can be done. But when we speak of a harsh economics of photons it would be quite wrong to assume that the harshness is confined to the night. In bright sunlight the photons may drum the retina like a tropical cloudburst, but there is still a problem. The essence of seeing a patterned image is that photocells in different parts of the retina must report different intensities of light and this means distinguishing different rates of pattering in different parts of the photon rainstorm. The sorting of photons coming from different fine-grained parts of the scene can lead to local impoverishments of photons just as serious as the global impoverishment of the night. It is to this sorting that we now turn.

Photocells on their own just tell an animal whether there is light or not. The animal can tell the difference between night and day, and can tell when a shadow falls which might, for example, portend a predator. The next step of improvement must have been the acquisition of some rudimentary sensitivity to direction of light and direction of movement of, say, a menacing shadow. The minimal way of achieving this is to back the photocells with a dark screen on one side only. A transparent photocell without a dark screen receives light from all directions and cannot tell where light is coming from. An animal with only one photocell in its head can steer towards, or away from, light, provided the photocell is backed by a screen. A simple recipe for doing this is to swing the head like a pendulum from side to side; if the light intensity on the two sides is unbalanced, change direction until it is balanced. There are some maggots that follow this recipe for steering directly away from light.

But swinging your head from side to side is a rudimentary way of detecting the direction of light, fit for the lowest slopes of Mount Improbable. A better way is to have more than one photocell pointing in different directions, each one backed by a dark screen. Then by comparing the rates of photon rain on the two cells you can make inferences about direction of light. If you have a whole carpet of photocells, a better way is to bend the carpet, with its backing screen, into a curve, so that the photocells on different parts of the curve are pointing in systematically different directions. A convex curve can give rise, eventually, to the sort of 'compound eye' that insects have, and I'll return to this. A concave curve is a cup and it gives rise to the other main kind of eye, the camera eye like our own. Photocells in different parts of a cup will fire when light is coming from different directions, and the more cells there are the finer-grained will be the discrimination.

The light rays (parallel white lines with arrows) are halted by the thick black screen at the back of the cup (Figure 5.3). By keeping track of which photocells are firing and which are not, the brain can detect the direction from which the light is coming. From the point of view of climbing Mount Improbable, what matters is that there is a continuous evolutionary gradation—a smooth incline up the moun-

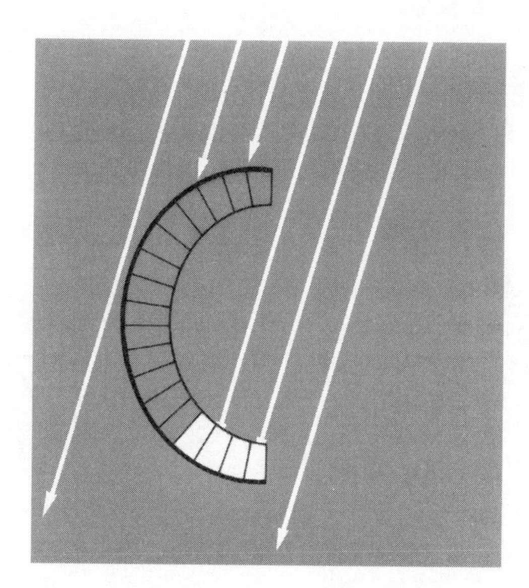

Figure 5.3 A simple cup eye can detect the direction of light.

tain—connecting animals with a flat carpet of photocells to animals with a cup. Cups can get gradually deeper or gradually shallower, by continuous slow degrees. The deeper the cup, the greater the ability of the eye to discriminate light coming from different directions. On the mountain, no steep precipices have to be leapt.

Cup eyes like this are common in the animal kingdom. Figure 5.4 shows the eye of a limpet, of a bristleworm, of a clam and of a flat-

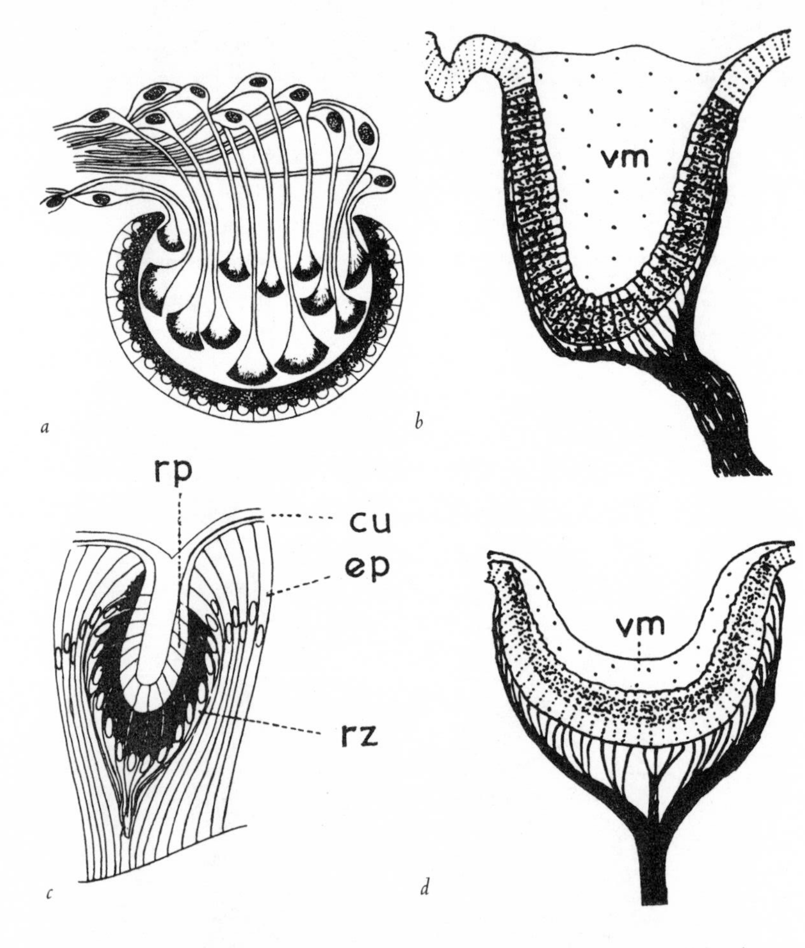

Figure 5.4 Cup eyes from around the animal kingdom: *(a)* flatworm; *(b)* bivalve mollusc; *(c)* polychaet worm; *(d)* limpet.

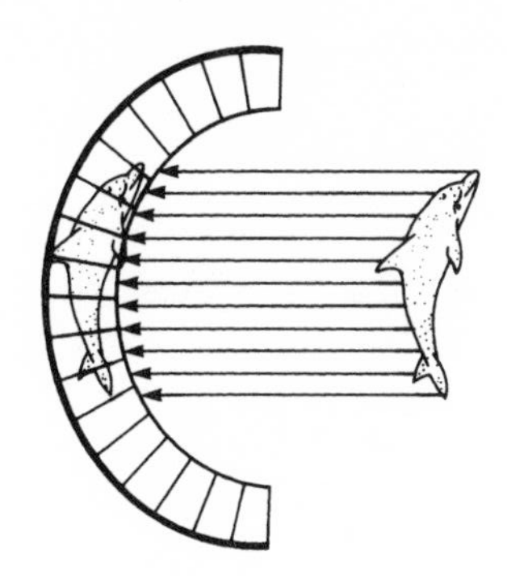

Figure 5.5 How eyes do not work—would that light rays were so obliging!

worm. These eyes have probably evolved their cup shape independently. This is particularly clear in the case of the flatworm eye which betrays its separate origin by keeping its photocells *inside* the cup. On the face of it this is an odd arrangement—the light rays have to penetrate a thicket of connecting nerves before they hit the photocells—but let's not be snobbish about it because the same apparently poor design mars our own much more sophisticated eyes. I'll return to this and show that it isn't really such a bad idea as it seems.

In any case, a cup eye on its own is far from capable of forming what we humans, with our excellent eyes, would recognize as a proper image. Our kind of image-formation, which depends upon the lens principle, needs a little explanation. We approach the problem by asking why an unaided carpet of photocells, or a shallow cup, will not see an image of, say, a dolphin, even when the dolphin is conspicuously displayed in front of it.

If light rays behaved as in Figure 5.5, everything would be easy and an image of the dolphin, the right way up, would appear on the retina. Unfortunately they don't. To be more precise, there are rays that do exactly what I have drawn in the picture. The trouble is that these are swamped by any number of rays going in every other direction at the same time. Every bit of the dolphin sends a ray to every point on the retina. And not just every bit of the dolphin, but every bit of the background and of everything else in the scene. You can think of the result as an infinite number of dolphin images, in every possible position on the surface of the cup and every possible way up and way round. But

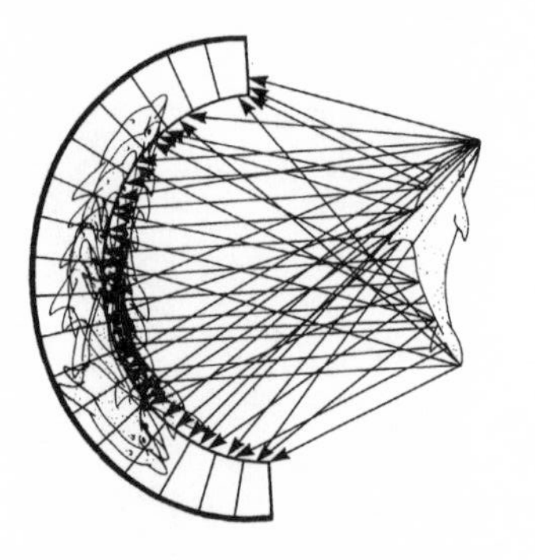

Figure 5.6 Light rays from everywhere go everywhere and no image is seen. An infinite number of dolphin images clash with each other, and nothing is clear.

what this amounts to, of course, is no image at all, just a smooth spreading of light over the whole surface (Figure 5.6).

We have diagnosed the problem. The eye is seeing too much: an infinity of dolphins instead of only one. The obvious solution is to subtract: cut out every dolphin image except one. It wouldn't matter which one, but how to get rid of the rest? One way is to trudge on up the same slope of Mount Improbable as gave us the cup, steadily deepening and enclosing the cup until the aperture has narrowed to a pinhole. Now the vast majority of rays are prevented from entering the cup. The minority that remain are just those rays that form a small number of similar images—upside-down—of the dolphin (Figure 5.7). If the pinhole becomes extremely small, the blurring disappears and a single, sharp picture of the dolphin remains (actually, extremely small pinholes introduce a new kind of blurring, but we'll forget about that for a moment). You can think of the pinhole as an image filter, removing all but one of the bewildering visual cacophony of dolphins.

The pinhole effect is just an extreme version of the cup effect that we have already met as an aid to telling the direction of light. It belongs only a bit farther up the same slope of Mount Improbable and there are no sharp precipices between. There is no difficulty in a pinhole eye's evolving from a cup eye, and no difficulty in a cup eye's evolving from a flat sheet of photocells. The slope up the mountain from flat carpet to pinhole is gradual and easily climbable all the way. Climbing it represents a progressive knocking

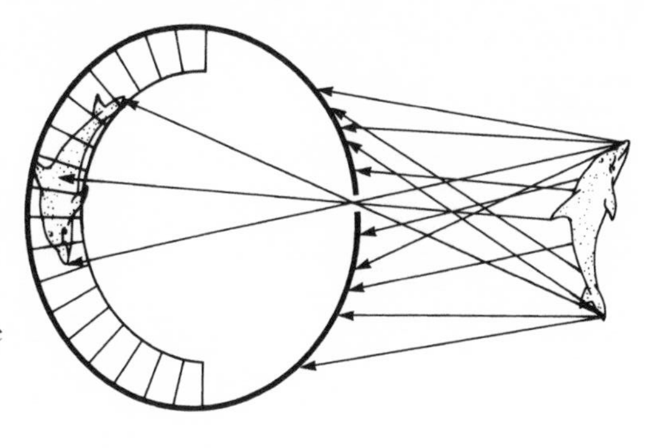

Figure 5.7 Principle of the pinhole eye. Most of the competing dolphin images are cut out. Ideally only one (inverted) gets through the pinhole.

out of conflicting images until, at the peak, only one is left.

Pinhole eyes in varying degrees are, indeed, scattered around the animal kingdom. The most thoroughgoing pinhole eye is that of the enigmatic mollusc *Nautilus* (Figure 5.8a), related to the extinct ammonites (and a more distant relative of an octopus, but with a coiled shell). Others, such as the eye of a marine snail in Figure 5.8b, are perhaps better described as deep cups than true pinholes. They illustrate the smoothness of this particular gradient up Mount Improbable.

A first thought suggests that the pinhole eye ought to work rather well, provided you make the pinhole small enough. If you make the pinhole almost infinitely small, you might think that you'd get an almost infinitely perfect image by cutting out the vast majority of competing, interfering images. But now two new snags arise. One is diffraction. I deferred talking about it just now. It is a blurring problem that results from the fact that light behaves like waves, which can interfere with each other. This blurring gets worse when the pinhole is very small. The other snag with a small pinhole recalls the hard trade-offs of our 'photon economy'. When the pinhole is small enough to make a sharp image, it necessarily follows that so little light gets through the hole that you can see the object well only if it is illuminated by an almost unattainably bright light. At normal lighting levels not enough photons get through the pinhole for the eye to be certain what it is seeing. With a tiny pinhole we have a version of the Halley's Comet problem. You can combat this by opening out the pinhole

Figure 5.8 A range of invertebrate eyes that illustrate approaches to the formation of crude but effective images: *(a) Nautilus*'s pinhole eye; *(b)* marine snail; *(c)* bivalve mollusc; *(d)* abalone; *(e)* ragworm.

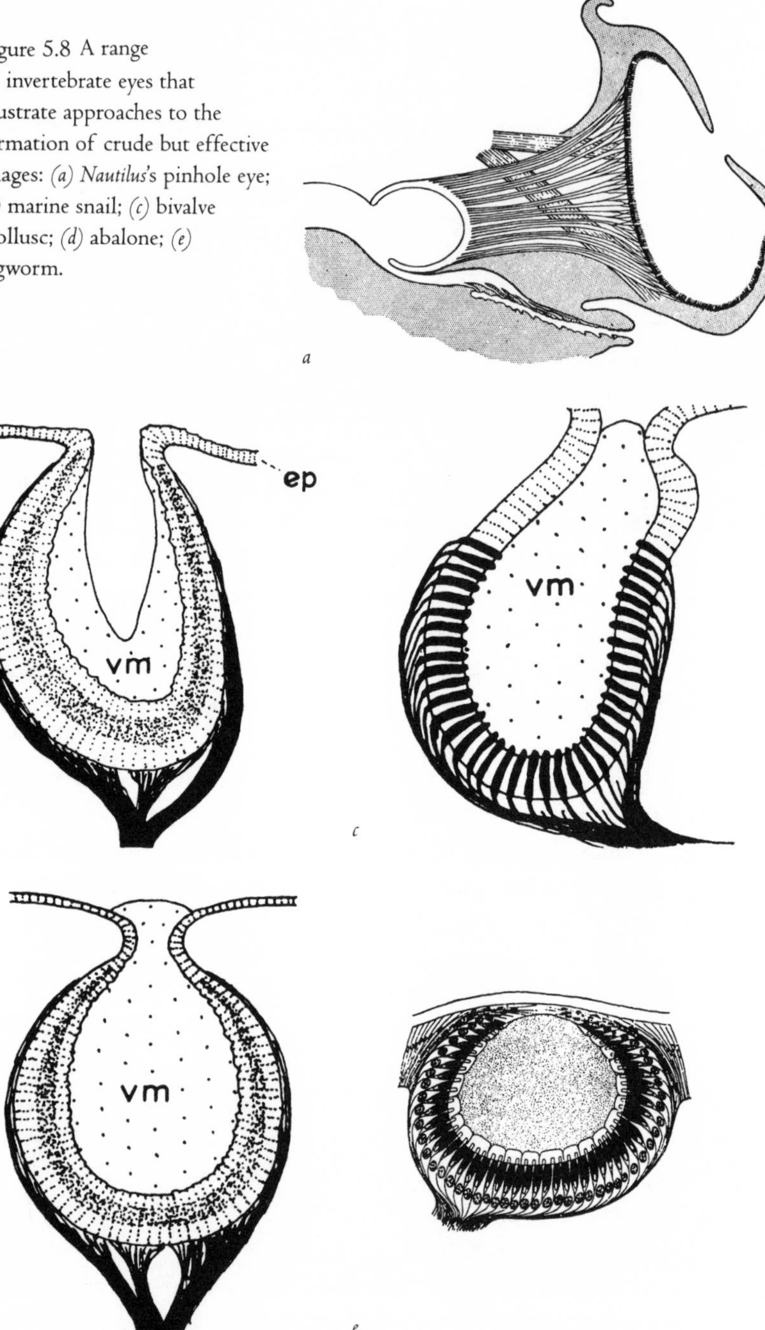

again. But now you are back where you were with a confusion of competing 'dolphins'. The photon economy has brought us to an impasse on this particular foothill of Mount Improbable. With the pinhole design you can have a sharpish but dark image, or a brightish but fuzzy one. You cannot have both. Such trade-offs are meat and drink to economists, which is why I coined the notion of an economy of photons. But is there no way to achieve a bright and yet simultaneously sharp image? Fortunately there is.

First, think of the problem computationally. Imagine that we broaden the pinhole out to let in a nice lot of light. But instead of leaving it as a gaping hole, we insert a 'magic window', a masterpiece of electronic wizardry embedded in glass and connected to a computer (Figure 5.9). The property of this computer-controlled window is the following. Light rays, instead of passing straight through the glass, are bent through a cunning angle. This angle is carefully calculated by the computer so that all rays originating from a point (say the dolphin's nose) are bent to converge on a corresponding point on the retina. I've drawn only the rays from the dolphin's nose, but the magic screen, of course, has no reason to favour any one point and does the same calculation for every other point as well. All rays originating at the tail of the dolphin are bent in such a way that they converge on a corresponding tail point on the retina, and so on. The

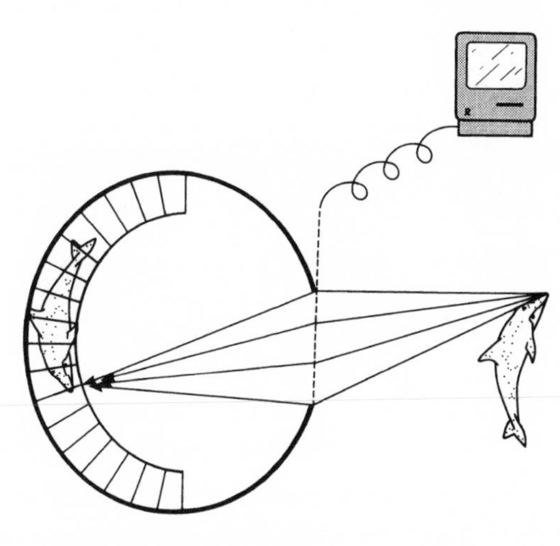

Figure 5.9 A complicated and ridiculously expensive hypothetical approach to the problem of forming an image that is both sharp and bright: the 'computed lens'.

result of the magic window is that a perfect image of the dolphin appears on the retina. But it is not dark like the image from a tiny pinhole, because lots of rays (which means torrents of photons) converge from the nose of the dolphin, lots of rays converge from the tail of the dolphin, lots of rays converge from every point on the dolphin to their own particular point on the retina. The magic window has the advantages of a pinhole, without its great disadvantage.

It's all very well to conjure up a so-called 'magic window' out of imaginative thin air. But isn't it easier said than done? Think what a complicated calculation the computer attached to the magic window is doing. It is accepting millions of light rays, coming from millions of points out in the world. Every point on the dolphin is sending millions of rays at millions of angles to different points on the surface of the magic window. The rays are criss-crossing one another in a bewildering spaghetti junction of straight lines. The magic window with its associated computer has to deal with each of these millions of rays in turn and calculate its own particular angle, through which it must be precisely turned. Where is this wonderful computer to come from, if not from a complicated miracle? Is this where we meet our Waterloo: an inevitable precipice in our ascent up Mount Improbable?

Remarkably, the answer is no. The computer in the diagram is just an imaginary creation to emphasize the *apparent* complexity of the task if you look at it in one way. But if you approach the problem in another way the solution turns out to be ludicrously easy. There is a device of preposterous simplicity which happens to have exactly the properties of our magic window, but with no computer, no electronic wizardry, no complication at all. That device is the lens. You don't need a computer because the calculations need never be done explicitly at all. The apparently complicated calculations of millions of ray angles are taken care of, automatically and without fuss, by a curved blob of transparent material. I'll take a little time to explain how lenses work, as a prelude to showing that the evolution of the lens wouldn't have been very difficult.

It is a fact of physics that light rays are bent when they pass from one transparent material into another transparent material (Figure

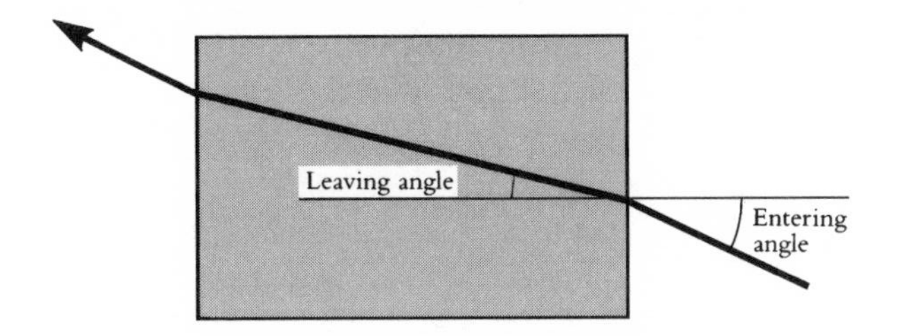

Figure 5.10 How light is bent. The principle of refraction in a block of glass.

5.10). The angle of bending depends upon which two materials they happen to be, because some substances have a greater refractive index—a measure of the capacity to bend light—than others. If we are talking about glass and water, the angle of bending is slight because the refractive index of water is nearly the same as that of glass. If the junction is between glass and air, the light is bent through a bigger angle because air has a relatively low refractive index. At the junction between water and air, the angle of bending is substantial enough to make an oar look bent.

Figure 5.10 represents a block of glass in air. The thick line is a light ray entering the block, being bent within the glass, then bending back to the original angle as it goes out the other side. But of course there is no reason why a blob of transparent material should have neatly parallel sides. Depending upon the angle of the surface of the blob, a ray can be sent off in any direction you choose. And if the blob is covered with facets at lots of different angles, a set of rays can be sent off in lots of different directions (Figure 5.11). If the blob is curved convexly on one or both of its sides, it will be a lens: the working equivalent of our magic window. Transparent materials are not particularly rare in nature. Air and water, two of the commonest substances on our planet, are both transparent. So are many other liquids. So are some crystals if their surface is polished, for instance by wave action in the sea, to remove surface roughness. Imagine a pebble of some crystalline material, worn into a random shape by the waves. Light rays from a single source are bent in all sorts of directions by the pebble, depending upon the angles of the pebble's surfaces.

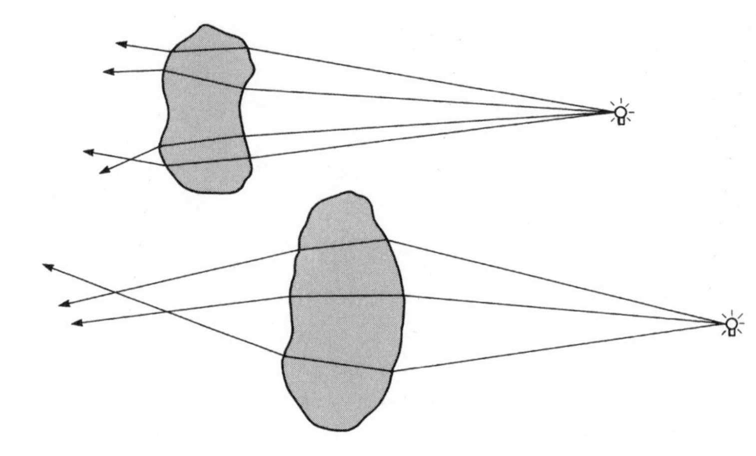

Figure 5.11 Random pebbles refract rays in unhelpful directions.

Pebbles come in all sorts of shapes. Quite commonly they are convex on both sides. What will this do to light rays from a particular source like a light bulb?

When the rays emerge from a pebble with vaguely convex sides they will tend to converge. Not to a neat, single point such as would reconstruct a perfect image of the light source like our hypothetical 'magic window'. That would be too much to hope. But there is a definite tendency in the right direction. Any quartz pebble whose weathering happened to make it smoothly curvaceous on both sides would serve as a good 'magic window', a true lens capable of forming images which, though far from sharp, are much brighter than a pinhole could produce. Pebbles worn by water usually are, as a matter of fact, convex on both sides. If they happened to be made of transparent material many of them would constitute quite serviceable, though crude, lenses.

A pebble is just one example of an accidental, undesigned object which can happen to work as a crude lens. There are others. A drop of water hanging from a leaf has curved edges. It can't help it. Automatically, without further design from us, it will function as a rudimentary lens. Liquids and gels fall automatically into curved shapes unless there is some force, such as gravity, positively opposing this. This will often mean that they cannot help functioning as lenses. The same is often true of biological materials. A young jellyfish is both lens-shaped and beautifully transparent. It works as a tolerably good lens, even though its lens properties are never actually used in life and there is no suggestion that natural selection has favoured its lens-like properties. The transparency probably is an advantage because it makes it hard for enemies to see, and the curved shape is an advantage for some structural reason having nothing to do with lenses.

Here are some images I projected on to a screen using various crude and undesigned image-forming devices. Figure 5.12a shows a large letter A, as projected on a sheet of paper at the back of a pinhole camera (a closed cardboard box with a hole in one side). You probably could scarcely read it if you weren't told what to expect, even though I used a very bright light to make the image. In order to get enough light to read it at all, I had to make the 'pin' hole quite large, about a centimetre across. I might have sharpened the image by narrowing the pinhole, but then the film would not have registered it—the familiar trade-off we have already discussed.

Now see what a difference even a crude and undesigned 'lens' makes. For Figure 5.12b the same letter A was again projected through the same hole on to the back wall of the same cardboard box. But this time I hung a polythene bag filled with water in front of

a

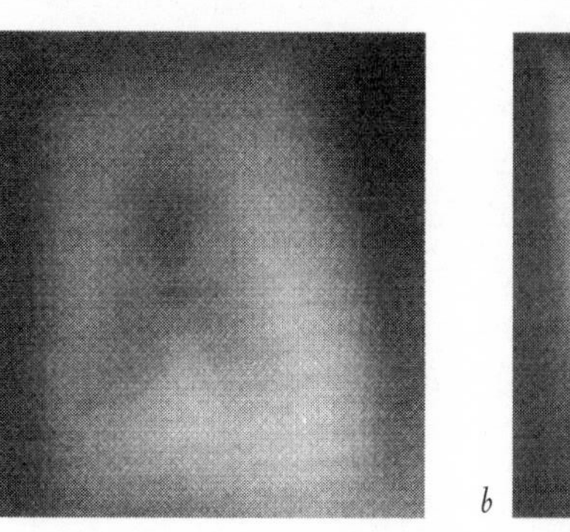

b

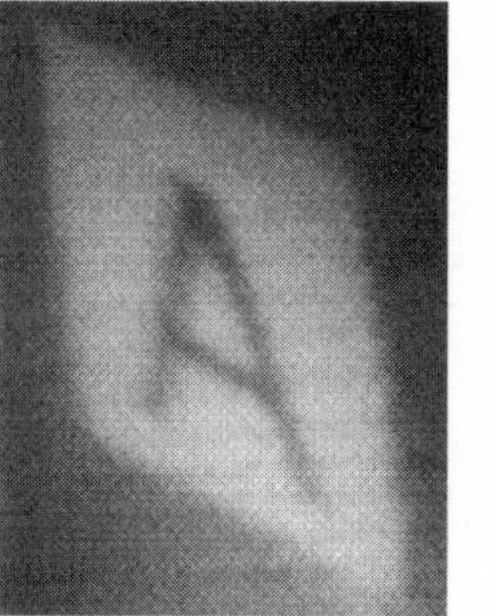

c

Figure 5.12 Images seen through the various makeshift holes and crude, makeshift lenses: (*a*) a plain pinhole; (*b*) a sagging polythene bag filled with water; (*c*) a round wine goblet filled with water.

the hole. The bag was not designed to be particularly lens-shaped. It just naturally hangs in a curvaceous shape when you fill it with water. I suspect that a jellyfish, being smoothly curved instead of rucked up into creases, would have produced an even better image. Figure 5.12c ('CAN YOU READ THIS?') was made with the same cardboard box and hole, but this time a round wine goblet filled with water was placed in front of the hole instead of a sagging bag. Admittedly the wineglass is a man-made object, but its designers never intended it to be a lens and they gave it its globular shape for other reasons. Once again, an object that was not designed for the purpose turns out to be an adequate lens.

Of course polythene bags and wineglasses were not available to ancestral animals. I am not suggesting that the evolution of the eye went through a polythene-bag stage, any more than it went through a cardboard-box stage. The point about the polythene bag is that, like a raindrop or a jellyfish or a rounded quartz crystal, it was not designed as a lens. It takes on a lens-like shape for some other reason which happens to be influential in nature.

It is not difficult, then, for rudimentary lens-like objects to come into existence spontaneously. Any old lump of half-way transparent jelly need only assume a curved shape (there are all sorts of reasons why it might) and it will immediately confer at least a slight improvement over a simple cup or pinhole. Slight improvement is all that is required to inch up the lower slopes of Mount Improbable. What might the intermediates have looked like? Look back at Figure 5.8, and once again I must stress that these animals are modern and must not be thought of as an actual ancestral series. Notice that the cup in Figure 5.8b (marine snail) has a lining of transparent jelly, the 'vitreous mass' (vm) which perhaps serves to protect the sensitive photocells from the raw sea water which flows freely through the aperture into the cup. That purely protective vitreous mass has one of the necessary qualities of a lens—transparency—but it lacks the correct curvature and it needs thickening up. Now look at Figures 5.8c, d and e, eyes from a bivalve mollusc, an abalone and a ragworm. In addition to providing yet more examples of cups and intermediates between cups and pinholes, all these eyes show greatly thickened vitreous masses. Vitreous masses, of varying degrees of shapelessness, are ubiquitous in the animal kingdom. As a lens, none of those splodges of jelly would move Mr Zeiss or Mr Nikon to write home. Nevertheless, any lump of jelly that has a little convex curvature would mark significant improvements over an open pinhole.

The biggest difference between a good lens and something like the abalone's vitreous mass is this: for best results the lens should be detached from the retina and separated from it by some distance. The gap need not be empty. It could be filled by more vitreous mass. What is needed is that the lens should have a higher refractive index than the substance that separates the lens from the retina. There are various ways in which this might be achieved, none of them difficult. I'll deal with just one way, in which the lens is condensed from a local region within the front portion of a vitreous mass like that in Figure 5.8e.

First, remember that a refractive index is something that every transparent substance has. It is a measure of its power to bend rays of light. Human lens-makers normally assume that the refractive index of a lump of glass is uniform through the glass. Once a ray of light has entered a particular glass lens and changed direction appropriately, it goes in a straight line until it hits the other side of the lens. The lens-maker's art lies in grinding and polishing the surface of the glass into precision shapes, and in joining different lenses together in compound cascades.

You can glue different kinds of glass together in complicated ways to make compound lenses with lots of different refractive indexes in various parts of them. The lens in Figure 5.13a, for instance, has a central core made of a different kind of glass with a higher refractive index. But there are still discrete changes from one refractive index to another. In principle, however, there is no reason why a lens should not have a continuously varying refractive index throughout its interior. This is illustrated in Figure 5.13b. This 'graded index lens' is hard for human lens-makers to achieve because of the way they make

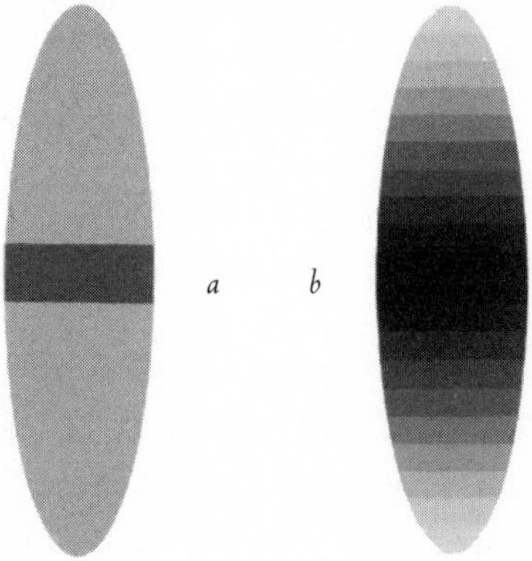

Figure 5.13 Two kinds of complex lens.

their lenses out of glass.* But it is easy for living lenses to be built like this because they are not made all at one time: they grow from small beginnings as the young animal develops. And, as a matter of fact, lenses with continuously varying refractive indexes are found in fish, octopuses and many other animals. If you look carefully at Figure 5.8e, you see what might conceivably be a region of differing refractive index in the zone behind the aperture of the eye.

But I was starting to tell the story of how lenses might have evolved in the first place, from a vitreous mass that filled the whole eye. The

*After writing this I was informed by a correspondent, Howard Kleyn, formerly of the Cable and Wireless Company, that humans do, as a matter of fact, make something equivalent to a graded index lens. It is actually a graded index optical fibre. By his description, it works like this. You start with a hollow tube of good glass, about a metre long and a few centimetres in diameter, which you heat up. You then puff into the tube finely powdered glass. The powdered glass melts and fuses with the lining of the tube, thereby thickening the lining while narrowing the bore of the tube. Now comes the cunning part. As this procedure progresses, the powder that is puffed in is of gradually changing quality: specifically, it has been ground from glass of progressively increasing refractive index. By the time the hollow bore has narrowed to nothing, the tube has become a rod made of highly refracting glass at its central core with gradually decreasing refractive index as you move towards its outer layers. The rod is then heated again, and drawn out into a fine filament. This filament retains the same graded refractive index, from core to periphery, in miniature, as the rod from which it was drawn. It is technically a graded index lens, albeit a very thin, long one. Its lens property is used not for focusing an image but for improving its quality as a light guide which does not allow its beam of light to disperse. Several of these filaments would normally be used to manufacture a multi-stranded optical fibre cable.

principle of how it might have happened, and the speed with which it might have been accomplished, has been beautifully demonstrated in a computer model by a pair of Swedish biologists called Dan Nilsson and Susanne Pelger. I shall lead up to explaining their elegant computer model in a slightly oblique way. Instead of going straight to what they actually did, I shall return to our progression from Biomorph to NetSpinner computer models and ask how one *could* ideally set about making a similar computer model of the evolution of an eye. I shall then explain that this is essentially equivalent to what Nilsson and Pelger did, although they didn't put it in quite the same way.

Recall that the biomorphs evolved by artificial selection: the selecting agent was human taste. We couldn't think of a realistic way of incorporating natural selection into the model so we switched to model spider webs instead. The advantage of spider webs was that, since they do their work in a two-dimensional plane, their efficiency in catching flies could be calculated by the computer automatically. So could their cost in silk, and model webs could therefore be automatically 'chosen' by the computer in a form of natural selection. We agreed that spider webs were exceptional in this respect: we could not easily hope to do the same for the backbone of a hunting cheetah or the fluke of a swimming whale, because the physical details involved in assessing a three-dimensional organ's efficiency are too complicated. But an eye is like a spider web in this respect. The efficiency of a model eye drawn in two dimensions can be assessed automatically by the computer. I am not implying that an eye is a two-dimensional structure, because it isn't. It is just that, if you assume that the eye is circular when seen head on, its efficiency in three dimensions can be assessed from a computer picture of a single vertical slice through the middle. The computer can do a simple ray-tracing analysis and work out the sharpness of image that an eye would be capable of forming. This quality scoring is equivalent to NetSpinner's calculation of the efficiency of a computer spider web at catching computer flies.

Just as NetSpinner webs procreated mutant daughter webs, so we could let model eyes generate mutant daughter eyes. Each daughter eye would have basically the same shape as the parent, but with a small random change to some minor aspect of its shape. Of course some of these computer 'eyes' would be so unlike real eyes as not to deserve the title, but no matter. They could still be bred, and their optical quality could still be given a numerical score—presumably it would be very low. We could therefore, in the same way as NetSpinner, evolve improved eyes by natural selection in the computer. We could either start with a fairly good eye and evolve a very good eye. Or we could start with a very poor eye or even with no eye at all.

It is instructive to run a program like NetSpinner as an actual simulation of evolution, setting it off from a rudimentary starting point and waiting to see where it will end up. You could even end up at different culmination points on different evolutionary runs, because there could be alternative accessible peaks of Mount Improbable. We could run our eye model in evolution mode too, and it would make a vivid demonstration. But actually you don't learn much more by letting the model evolve than you would learn by exploring, more systematically, where the upward path(s) on Mount Improbable lead(s). From a given starting point, a path which goes ever upward, never downward, is the path that natural selection would follow. If you ran the model in evolutionary mode, natural selection would follow that path. So it saves computer time if we search systematically for upward paths and for peaks that can be reached from postulated starting points. The important thing is that the rules of the game forbid going downhill. This more systematic search for upward paths is what Nilsson and Pelger did, but you can see why I chose to introduce their work *as if* we were planning, with them, a NetSpinner-style enactment of evolution.

However we choose to run our model, whether in 'natural-selection mode' or in 'systematic exploration of the mountain mode', we have to decide upon some rules of embryology: that is, some rules governing how genes control the development of bodies. What aspects of shape do the mutations actually operate upon? And how big, or how small, are the mutations themselves? In the case of NetSpinner, the mutations act upon known aspects of the behaviour of spiders. In the case of biomorphs, mutations act upon the lengths and angles of branches in growing trees. In the case of eyes, Nilsson and Pelger began by acknowledging that there are three main types of tissue in a typical 'camera' eye. There is an outer casing to the camera, usually opaque to light. There is a layer of light-sensitive 'photocells'. And there is some kind of transparent material, which may serve as a protective window or which may fill the cavity inside the cup—if, indeed, there is a cup, for we are not taking anything for granted in our simulation. Nilsson and Pelger's starting point—the foot of the mountain—is a flat layer of photocells (grey in Figure 5.14), sitting on a flat backing screen (black) and topped by a flat layer of transparent tissue (off-white). They assumed that mutation works by causing a small percentage change in the size of something, for example a small percentage decrease in the thickness of the transparent layer, or a small percentage increase in the refractive index of a local region of the transparent layer. Their question really is, where can you get to on the mountain if you start from a given base camp and go steadily upwards? Going upwards means mutating, one small step at a time, and only accepting mutations that improve optical performance.

So, where do we get to? Pleasingly, through a smooth upward pathway, starting from no proper eye at all, we reach a familiar fish eye, complete with lens. The lens is not uniform like an ordinary man-made lens. It is a graded index lens such as we met in Figure 5.13b. Its continuously varying refractive index is represented in the diagram by varying shades of grey. The lens has 'condensed' out of the vitreous mass by gradual, point by point changes in the refractive index. There is no sleight of hand here. Nilsson and Pelger didn't pre-program their simulated vitreous mass with a primordial lens just waiting to burst forth. They simply allowed the refractive index of each small bit of transparent material to vary under genetic control. Every smidgen of transparent material was free to vary its refractive index in any direction at random. An infinite number of patterns of varying refractive index *could* have emerged within the vitreous mass. What made the lens come out 'lens-shaped' was unbroken upward mobility, the equivalent of selectively breeding from the best seeing eye in each generation.

Nilsson and Pelger's purpose was not only to show that there is a smooth trajectory of improvement from a flat non-eye to a good fish

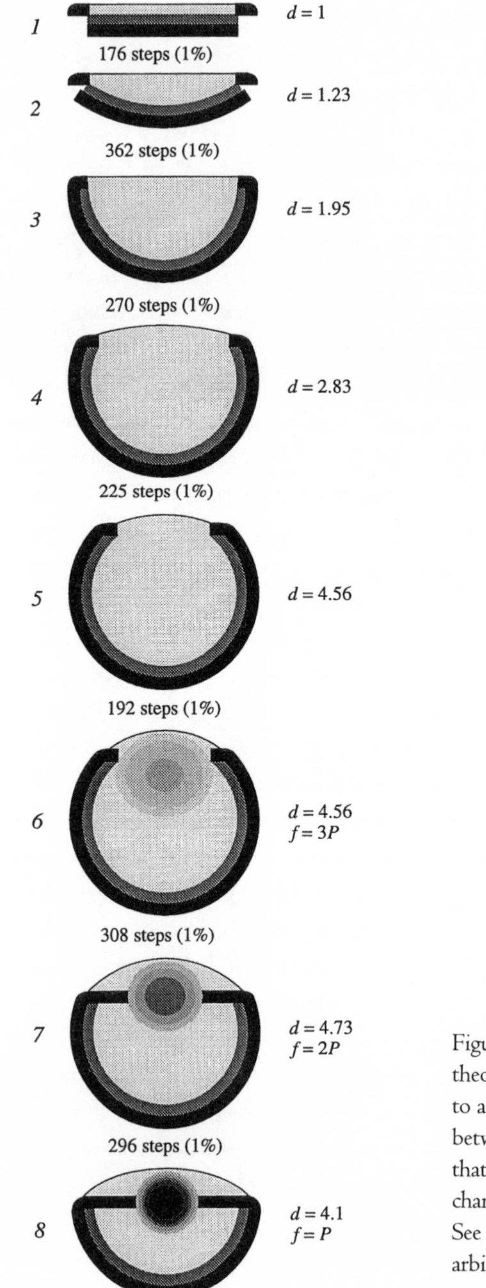

Figure 5.14 Nilsson and Pelger's theoretical evolutionary series leading to a 'fish' eye. The number of steps between stages assumes, arbitrarily, that each step represents a 1 per cent change in magnitude of something. See text for translation from these arbitrary units into numbers of generations of evolution.

eye. They were also able to use their model to estimate the time it would take to evolve an eye from nothing. The total number of steps that their model took was 1,829 if each step achieved a 1 per cent change in the magnitude of something. But there is nothing magic about 1 per cent. The same total quantity of change would have taken 363,992 steps of 0.005 per cent. Nilsson and Pelger had to re-express the total quantity of change in non-arbitrary, realistic units, and that means units of genetic change. In order to do this, it was necessary to make some assumptions. For example they had to make an assumption about the intensity of selection. They assumed that for every 101 animals possessing an improved eye who survived, 100 animals without the improvement survived. As you can see, this is a low intensity of selection as common sense might judge it—you are almost as well off without the improvement as with it. They deliberately chose a low, conservative or 'pessimistic' figure because they were bending over backwards to bias their estimate of rate of evolution towards being, if anything, too slow. They also had to make two other assumptions: of 'heritability' and of 'coefficient of variation'. The coefficient of variation is a measure of how much variation there is in the population. Natural selection needs variation to work upon and Nilsson and Pelger again deliberately chose a pessimistically low value. Heritability is a measure of how much of the variation, out of a given population's available variation, is inherited. If the heritability is low it means that most of the variation in the population is environmentally caused, and natural selection, for all that it may 'choose' individuals to live or die, will have little impact on evolution. If heritability is high, selection has a large impact on future generations because individual survival really translates into gene survival. Heritabilities frequently turn out to be more than 50 per cent, so Nilsson and Pelger's decision to settle on 50 per cent was a pessimistic assumption. Finally, they made the pessimistic assumption that different parts of the eye could not change simultaneously in one generation.

'Pessimistic' in all these cases means that the estimate that we finally come up with for how long it takes an eye to evolve is likely to be on the long side. The reason we call an over-estimate pessimistic rather than optimistic is this. A sceptic about the power of evolution, such as Emma Darwin, is naturally drawn to the view that an organ as notoriously complicated and many-parted as an eye, if it can evolve at all, will take an immense time to evolve. Nilsson and Pelger's final estimate was actually astoundingly short. At the end of the calculation, it turned out that it would take only about 364,000 generations to evolve a good fish eye with a lens. It would have been even shorter if they had made more optimistic (and this probably means more realistic) assumptions.

How long is 364,000 generations in years? That depends on the generation time, of course. The animals we are talking about would be small marine animals like worms, molluscs and small fish. For them, a generation typically takes one year or less. So Nilsson and Pelger's conclusion is that the evolution of the lens eye could have been accomplished in less than half a million years. And that is a very very short time indeed, by geological standards. It is so short that, in the strata of the ancient eras we are talking about, it would be indistinguishable from instantaneous. The plaint that there hasn't been enough time for the eye to evolve turns out to be not just wrong but dramatically, decisively, ignominiously wrong.

Of course there are some other details of a full-fledged eye that Nilsson and Pelger have not yet dealt with and which might (though they don't think so) take rather longer to evolve. There is the preliminary evolution of the light-sensitive cells—what I have been calling photocells—which they regarded as having been accomplished before the start of their model evolution system. There are other, advanced features of modern eyes such as the apparatuses for changing the focus of an eye, for changing the size of the pupil or '*f*-stop', and for moving the eye. There are also all the systems in the brain that are needed for processing the information from the eye. Moving the eye is important, not just for the obvious reason but, more indispensably, to hold the gaze still while the body moves. Birds do this by using the neck muscles to keep the whole head still, notwithstanding substantial movements of the rest of the body. Advanced systems for doing this involve quite sophisticated brain mechanisms. But it is easy to see that rudimentary, imperfect adjustments would be better than nothing, so there is no difficulty in piecing together an ancestral series following a smooth path up Mount Improbable.

In order to focus rays that are coming from a very distant target, you need a weaker lens than to focus rays that are coming from a close target. To focus sharply both far and near is a luxury one can live without, but in nature every little boost to the chances of survival counts and as a matter of fact different sorts of animals display a variety of mechanisms for changing the focus of the lens. We mammals do it by means of muscles that pull on the lens and change its shape a little. So do birds and most reptiles. Chameleons, snakes, fishes and frogs do it in the same way a camera does, by pulling the lens a little way forwards or backwards. Animals with smaller eyes don't bother. Their eyes are like a Box Brownie: approximately, though not brilliantly, in focus at all distances. As we get older our eyes sadly become more Box Brownie-like and we often need bifocal glasses to see both near and far.

It is not at all difficult to imagine the gradual evolution of mechanisms for changing focus. When experimenting with the polythene bag filled with water, I quickly noticed that the sharpness of focus could be made better (or worse) by poking the bag with my fingers. Without being consciously aware of the shape of the bag, without even looking at the bag but concentrating on the quality of the image being projected, I simply poked and squashed the bag at random until the focus got better. Any muscle in the vicinity of a lump of vitreous mass could, as a by-product of contracting for some other purpose, incidentally improve the focus of the lens. This opens up a broad highway for gentle improvement all the way up the slopes of Mount Improbable, which could culminate in either the mammal or the chameleon method of changing the focus.

Changing the aperture—the size of the hole through which light is admitted—may be slightly more difficult, but not much. The reason for wanting to do this is the same as in a camera. For any given sensitivity of film/photocells, it is possible to have too much light (dazzle) as well as too little. Moreover, the narrower the hole, the better the depth of focus—the range of distances that are simultaneously in focus. A sophisticated camera, or eye, has a built-in light meter which automatically stops down the hole when the sun comes out, and opens up the hole when the sun goes in. The pupil of a human eye is a pretty sophisticated piece of automation technology, something that a Japanese micro-engineer could be proud of.

But, once again, it isn't difficult to see how this advanced mechanism might have got its start on the lower slopes of Mount Improbable. We think of the pupil as circular, but it doesn't have to be. Any shape would do. Sheep and cattle have a long, horizontal, lozenge-shaped pupil. So do octopuses and some snakes, but other snakes have a vertical slit. Cats have a pupil which varies from a circle to a narrow, vertical slit (Figure 5.15):

> Does Minnaloushe know that his pupils
> Will pass from change to change,
> And that from round to crescent,
> From crescent to round they range?
> Minnaloushe creeps through the grass
> Alone, important and wise,
> And lifts to the changing moon
> His changing eyes.
>
> W. B. Yeats

Even expensive cameras often have pupils which are crude polygons rather than perfect circles. All that matters is that the quantity of light entering the eye should be controlled. When you realize this, the early evolution of the variable pupil ceases to be a problem. There are lots of gentle paths to be followed up the lower slopes of Mount Improbable. The iris diaphragm is no more an impenetrable evolutionary barrier than is the anal sphincter. Perhaps the most important quantity that needs to be improved is the speed of responsiveness of the pupil. Once you have nerves at all, speeding them up is an easy glide up the slopes of the mountain. Human pupils respond fast, as you can quickly verify by shining a torch in your eye while looking at your pupil in a mirror. (You see the effect most dramatically if you shine the torch in one eye while looking at the pupil in the other: for the two are ganged together.)

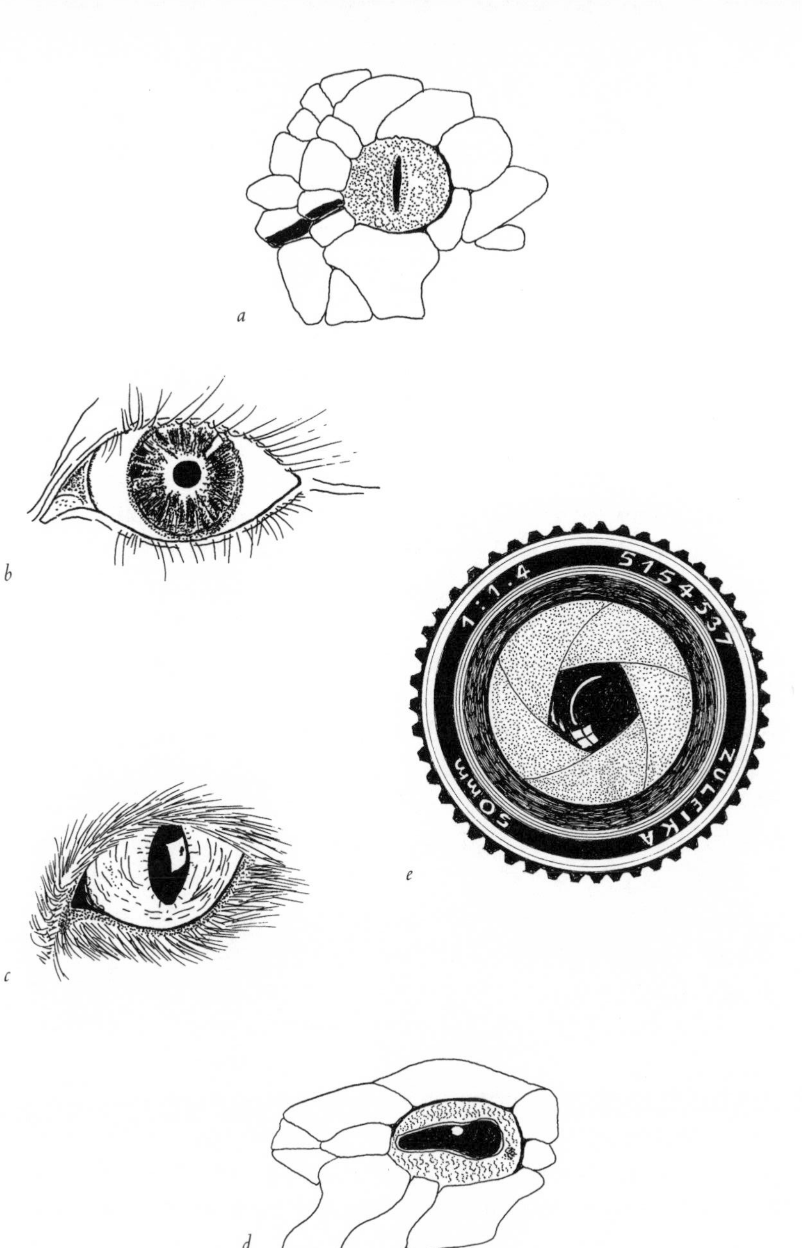

Figure 5.15 Various pupils including that of a camera. The exact shape of a pupil doesn't matter, which is why it is allowed to be so variable: *(a)* reticulated python; *(b)* human; *(c)* cat; *(d)* long-nosed tree snake; *(e)* camera.

As we've seen, the Nilsson and Pelger model developed a graded index lens, which is different from most man-made lenses but like those of fishes, squids and other underwater camera eyes. The lens arises by condensation of a zone of locally high-refractive index within previously uniform transparent jelly.

Not all lenses evolved by condensing out from a gelatinous mass. Figure 5.16 shows two insect eyes that form their lenses in quite different ways. These are both so-called simple eyes, not to be confused with the compound eyes which we'll come to in a moment. In the first of these simple eyes, from a sawfly larva, the lens forms as a thickening of the cornea—the outer transparent layer. In the second one, from a mayfly, the cornea is not thickened and the lens develops as a mass of colourless, transparent cells. Both these two methods of lens development lend themselves to the same kind of Mount Improbable climb as we've already undertaken for the vitreous mass eye of the worm. Lenses, like eyes themselves, seem to have evolved many times independently. Mount Improbable has many peaks and hillocks.

Retinas, too, betray their manifold origins by their variable forms. With one exception, all the eyes I have so far illustrated have had their photocells in front of the nerves connecting them to the brain. This is the obvious way to do it, but it is not universal. The flatworm in Figure 5.4a keeps its photocells apparently on the wrong side of their connecting nerves. So does our own vertebrate eye. The photocells point backwards, away from the light. This is not as silly as it sounds. Since they are very tiny and transparent, it doesn't much matter which way they point: most photons will go straight through and then run the gauntlet of pigment-laden baffles waiting to catch them. The only sense in which it even means much to say that vertebrate photocells point backwards is that the 'wires' (nerves) connecting them to the brain depart in the wrong direction, towards the light rather than towards the brain. They then run over the front surface of the retina towards one particular place, the so-called 'blind spot'. This is where they dive through the retina into the optic nerve, which is why the retina is blind at this spot. Although we are all technically blind at the spot, we scarcely know it because the brain is so clever at reconstituting the missing bit. We only notice the blind spot if the

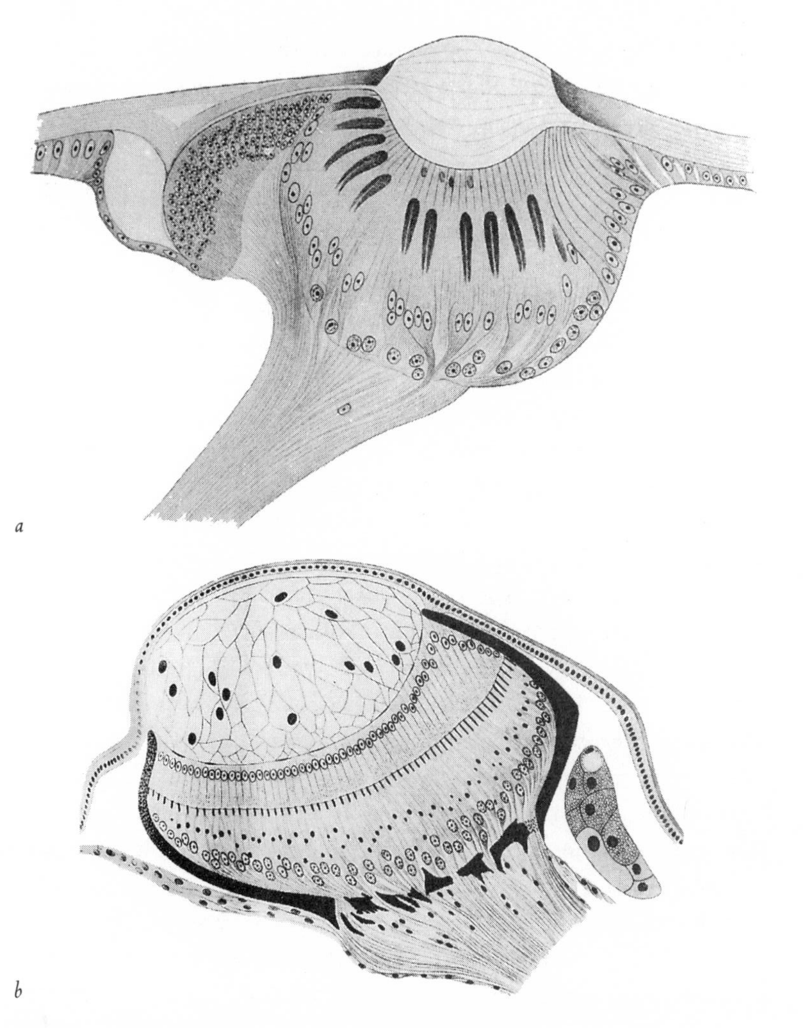

Figure 5.16 Two different ways for insect lenses to develop: *(a)* sawfly larva; *(b)* mayfly.

image of some small discrete object, which we have independent evidence exists, moves on to it: it then appears to go out like a light, apparently replaced by the general background colour of the area.

I've said that it makes little difference if the retina is back-to-front. A case could be made that, absolutely all other things being equal, it might have been better if our retinas were the right way round. It is perhaps a good example of the fact that Mount Improbable has more than one peak, with deep valleys between. Once a good eye has started to evolve with its retina back-to-front, the only way to ascend is to improve the present design of eye. Changing to a radically different design involves going downhill, not just a little way but down a deep chasm, and that is not allowed by natural selection. The vertebrate retina faces the way it does because of the way it develops in the embryo, and this certainly goes back to its ancient ancestors. The eyes of many invertebrates develop in different ways, and their retinas are consequently the 'right' way round.

Setting aside the interesting fact of their pointing backwards, vertebrate retinas scale some of the loftiest peaks on the mountain. The human retina has about 166 million photocells, divided into various kinds. The basic division is into rods (specialized for low-precision, non-colour vision at relatively low light levels) and cones (specialized for high-precision colour vision in bright light). As you read these words you are using only cones. If Juliet had seen Halley's Comet, it would have been her rods that were responsible. The cones are concentrated in a small central area, the fovea (you are reading with your foveas) where there are no rods. This is why, if you want to see a really dim object like Halley's Comet, you must point your eyes not directly at it but slightly away, so that its meagre light is off the fovea. Numbers of photocells, and differentiation of photocells into more than one type, present no special problems from the point of view of climbing Mount Improbable. Both kinds of improvement obviously constitute smooth gradients up the mountain.

Big retinas see better than small retinas. You can fit more photocells in, and you can see more detail. But, as always, there are costs. Remember the surrealist snail of Figure 5.1. But there is a way in which a small animal can, in effect, enjoy a larger retina than it pays for. Professor Michael Land of Sussex University, who has an enviable track record for exotic discoveries in the world of eyes and from whom I have learned much of what I know about eyes, found a wonderful example in jumping spiders.* No spiders have compound eyes: jumping spiders have taken the camera eye up to a remarkable peak of economy (Figure 5.17). What Land discovered was an extraordinary retina. Instead of being a wide sheet on which a full image can be projected, it is a long, vertical strip, not wide enough to accommodate a decent image. But the spider makes up for the narrowness of its retina by an ingenious makeshift. It moves its retina systematically about, 'scanning' the area where an image might be projected. Its effective retina is, therefore, much larger than its actual retina—rather on the same principle as the bolas spider with its whirling single thread approximates the catchment area of a proper web. If the jumping spider's retina finds an interesting object, such as a moving fly or another jumping spider, it concentrates its scanning movements in the precise area of the target. This gives it the dynamic equivalent of a fovea. Using this clever trick, jumping spiders have carried the lens eye to a respectable little peak in their local area of Mount Improbable.

I introduced the lens as an excellent remedy for the shortcomings of the pinhole eye. It isn't the only one. A curved mirror constitutes a different principle from a lens, but it is a good alternative solution to the same problem of gathering a large amount of light from each point on an object, and focusing it to a single point on an image. For some purposes a curved mirror is actually a more economical solution to the problem than a lens, and the biggest optical telescopes in the world are all reflectors (Figure 5.18a). A minor problem with a reflecting telescope is that the image is formed in front of the mirror, actually in the pathway of the incoming rays. Reflecting telescopes usually have a small mirror to reflect the focused image sideways into an eyepiece or a camera. The small mirror doesn't get in the way, not

*These engaging little animals, whose habit of cocking their heads to look at you gives them an almost human charm, stalk their prey like a cat and then jump on to it explosively and without warning. Explosive it more or less literally is, by the way, for they jump by hydraulically pumping fluid into all eight legs simultaneously—a little like the way we (those of us who have them) erect our penises, but their 'leg erections' are sudden rather than gradual.

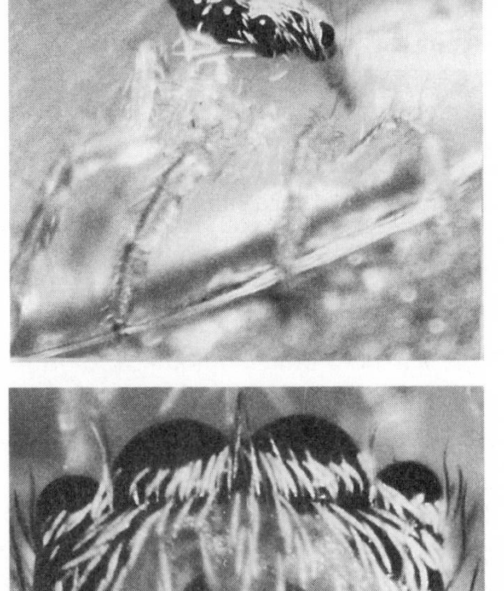

Figure 5.17 Jumping spider.

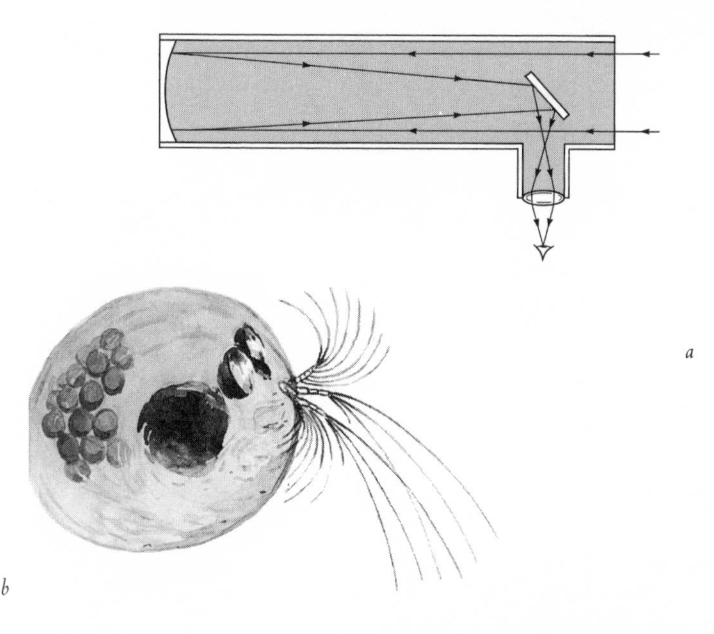

Figure 5.18 Curved mirror solutions to the problem of forming images: *(a)* reflecting telescope; *(b)* *Gigantocypris,* a large planktonic crustacean painted by Sir Alister Hardy; *(c)* scallop eyes peeping through gap in shell; *(d)* cross-section of scallop eye; *(e)* Cartesian oval.

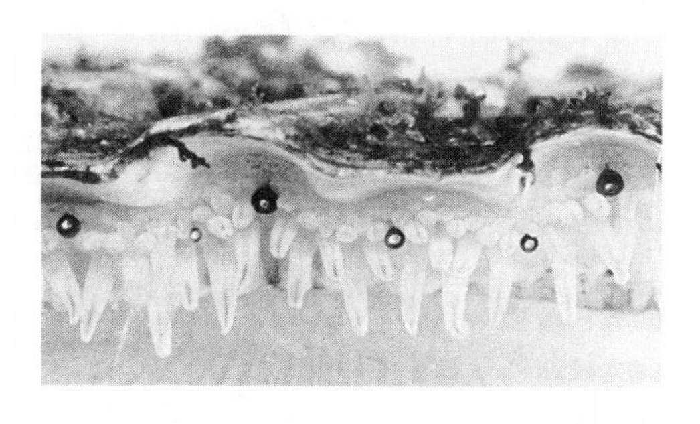

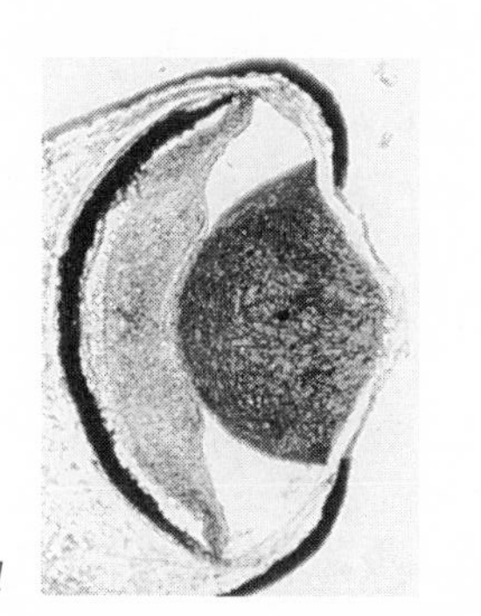

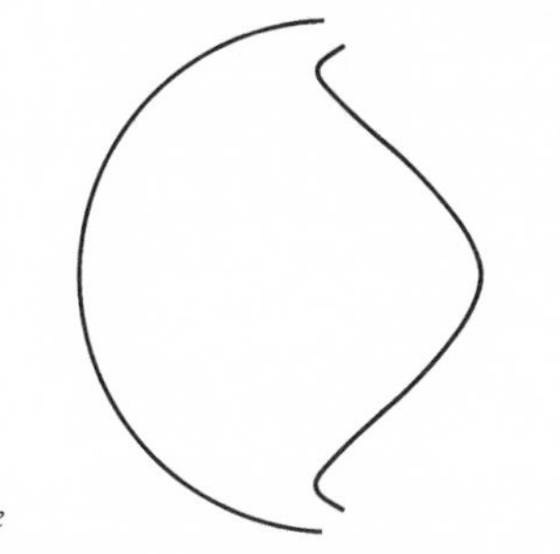

enough to spoil the image, anyway. No focused image of the little mirror is seen: it merely causes a small reduction in the total amount of light hitting the big mirror at the back of the telescope.

The curved mirror, then, is a theoretically workable physical solution to an important problem. Are there any examples of curved mirror eyes in the animal kingdom? The earliest suggestion along these lines was made by my old Oxford Professor, Sir Alister Hardy, commenting on his painting of a remarkable deep-sea crustacean called *Gigantocypris* (Figure 5.18b). Astronomers capture what few photons arrive from distant stars with huge curved mirrors in observatories like Mount Wilson and Palomar. It is tempting to think that *Gigantocypris* is doing the same thing with the few photons that penetrate the deep oceans, but recent investigations by Michael Land rule out any resemblance in detail. It is at the moment not clear how *Gigantocypris* sees.

There is another kind of animal, however, that definitely uses a bona fide curved mirror to form an image, albeit it has a lens to help. Once again, it was discovered by that King Midas of animal eye research, Michael Land. The animal is the scallop.

The photograph in Figure 5.18c is an enlargement of a small piece (two shell-corrugations in width) of the gap of one of these bivalves. Between the shell and the tentacles is a row of dozens of little eyes. Each eye forms an image, using a curved mirror which lies well behind the retina. It is this mirror that causes each eye to glow like a tiny blue or green pearl. In section, the eye looks like Figure 5.18d. As I mentioned, there is a lens as well as a mirror, and I'll come back to this. The retina is the whole greyish area lying between the lens and the curved mirror. The part of the retina which sees the sharp image projected by the mirror is the portion tightly abutting the back of the lens. That image is upside-down and it is formed by rays reflected backwards by the mirror.

So, why is there a lens at all? Spherical mirrors like this one are subject to a particular kind of distortion called spherical aberration. A famous design of reflecting telescope, the Schmidt, overcomes the problem by a cunning combination of lens and mirror. Scallop eyes seem to solve the problem in a slightly different way. Spherical aberration can theoretically be overcome by a special kind of lens whose shape is called a 'Cartesian oval'. Figure 5.18e is a diagram of a theoretically ideal Cartesian oval. Now look again at the profile of the actual lens of the scallop eye (Figure 5.18d). On the basis of the striking resemblance, Professor Land suggests that the lens is there as a corrector for the spherical aberration of the mirror which is the main image-forming device.

As for the origin of the curved mirror eye on the lower slopes of its region of the mountain, we can make an educated guess. Reflecting layers behind retinas are common in the animal kingdom, but for a different purpose, not image forming, as in scallops. If you go out into the woods with a bright spotlight you will see numerous twin beams glaring straight back at you. Many mammals, especially nocturnal ones like Figure 5.19b's golden potto or angwantibo from West Africa, have a tapetum, a reflecting layer behind the retina. What the tapetum does is provide a second chance of catching photons that the photocells failed to stop: each photon is reflected straight back to the very photocell that missed it coming the other way, so the image is not distorted. Invertebrates, too, have discovered the tapetum. A bright torch in the woods is an excellent way to find certain kinds of spider. Indeed, looking at the portrait of a wolf spider (Figure 5.19a), you wonder why the 'cats' eyes' that mark our roads are not called 'spider eyes'. Tapetums for capturing every last photon may well have evolved in ancestral cup eyes before lenses. Perhaps the tapetum is the pre-adaptation which, in a few isolated creatures, has become modified to form a reflecting telescope kind of eye. Or the mirror may have arisen from another source. It is hard to be sure.

Lenses and curved mirrors are two ways of sharply focusing an image. In both cases the image is upside-down and left–right reversed. A completely different kind of eye, which produces an image the right way up, is the compound eye, favoured by insects, crustaceans, some worms and molluscs, king crabs (strange marine creatures said to be closer to spiders than to real crabs) and the large group of now extinct trilobites. Actually there are several different kinds of compound eye. I'll begin with the most elementary kind, the so-called apposition compound eye. To understand how the apposition eye

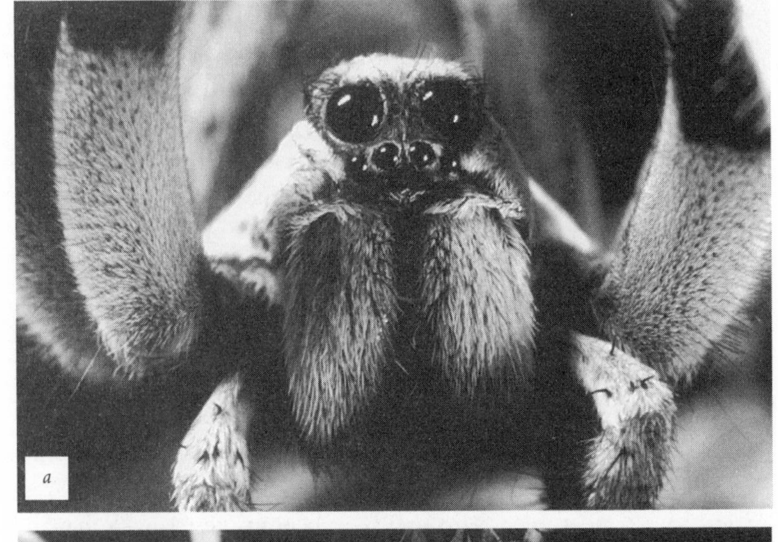

Figure 5.19 Saving photons by reflecting them back. Glowing tapeta behind the eyes of *(a)* a wolf spider, *Geolycosa sp.*, and *(b)* a golden potto.

works, we go back nearly to the bottom of Mount Improbable. As we have seen, if you want an eye to see an image or indeed go beyond signalling the mere intensity of light, you need more than one photocell, and they must pick up light from different directions. One way to make them look in different directions is to place them in a cup, backed by an opaque screen. All the eyes we have so far talked about have been descendants of this concave cup principle. But perhaps an even more obvious solution to the problem is to place the photocells on the convex, outside surface of a cup, thereby causing them to look outwards in different directions. This is a good way to think of a compound eye, at its simplest.

Remember when we first introduced the problem of forming an image of a dolphin. I pointed out that the problem could be regarded as the problem of having too many images. An infinite number of 'dolphins' on the retina, every way up and in every position on the retina, adds up to no visible dolphin at all (Figure 5.20a). The pinhole eye worked because it filtered out almost all the rays, leaving only the minority that cross each other in the pinhole and form a single upside-down image of the dolphin. We treated the lens as a more sophisticated version of the same principle. The ap-

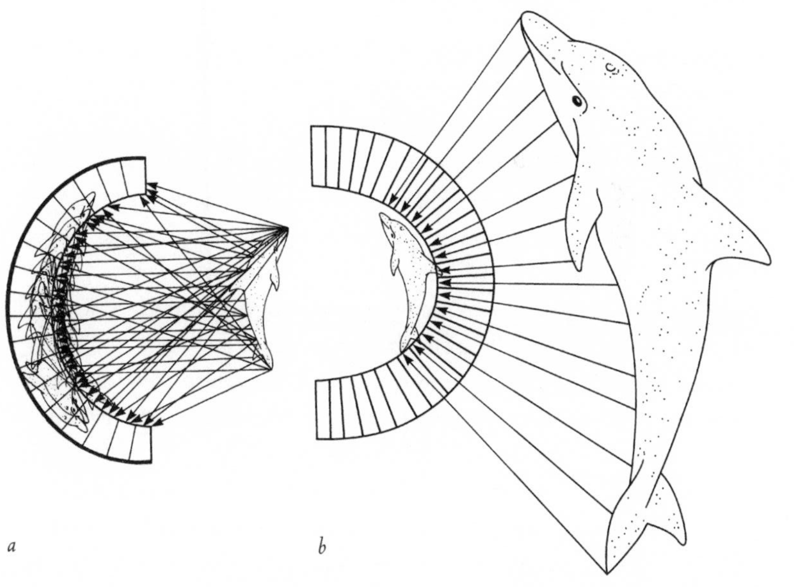

Figure 5.20 *(a)* reproduction of Figure 5.6; *(b)* the cup turned inside out. Principle of the apposition compound eye.

position compound eye solves the problem in an even simpler way.

The eye is built as a dense cluster of long straight tubes, radiating out in all directions from the roof of a dome. Each tube is like a gunsight which sees only the small part of the world in its own direct line of fire. In terms of our filtering metaphor, we could say that rays coming from other parts of the world are prevented, by the walls of the tube and the backing of the dome, from hitting the back of the tube where the photocells are.

That's basically how the apposition compound eye works. In practice, each of the little tube eyes, called an ommatidium (plural ommatidia), is a bit more than a tube. It has its own private lens, and its own tiny 'retina' of, usually, half a dozen or so photocells. Insofar as each ommatidium produces an image at all at the bottom of the narrow tube, that image is upside-down: the ommatidium works like a long, poor-quality, camera eye. But the upside-down images of the individual ommatidia are ignored. The ommatidium reports only how much light comes down its tube. The lens serves only to gather more light rays from the ommatidium's gunsight direction and focus them on to the retina. When all the ommatidia are taken together, their summed 'image' is the right way up, as shown in Figure 5.20b.

As always, 'image' doesn't have to mean what we humans would think of as an image: an accurate, Technicolor perception of an entire scene. Instead, we are talking about *any* kind of ability to use the eyes to distinguish what is going on in different directions. Some insects might, for example, use their compound eyes only to track moving targets. They might be blind to still scenes. The question of whether animals see things in the same way as we do is partly a philosophical one and it may be a more than usually difficult task trying to answer it.

The compound-eye principle works well enough for, say, a dragonfly zeroing in on a moving fly but, in order for a compound eye to see as much detail as we see, it would need to be hugely bigger than the kind of simple camera eye that we possess. Here is approximately why this is. Obviously, the more ommatidia you have, all looking in slightly different directions, the more fine detail you can see. A dragonfly may have 30,000 ommatidia and it is pretty good at hawking insects on the wing (Figure 5.21). But in order to see as much detail

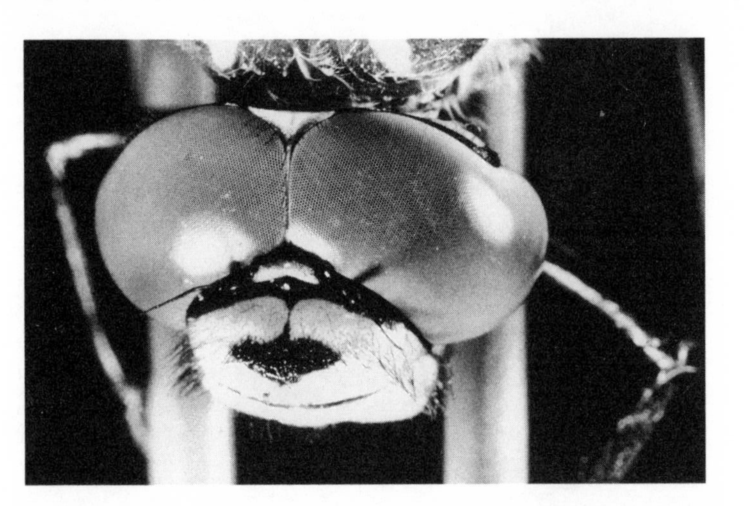

Figure 5.21 Large compound eyes in a visually hunting aerial predator, the dragonfly *Aeshna cyanea.*

as we can see, you'd need millions of ommatidia. The only way to fit in millions of ommatidia is to make them exceedingly tiny. And unfortunately there is a strict limit on how small an ommatidium can be. It is the same limit as we met in talking about very small pinholes, and it is called the diffraction limit. The consequence is that, in order to make a compound eye see as precisely as the human camera eye, the compound eye would have to be ludicrously large: twenty-four metres in diameter! The German scientist Kuno Kirschfeld dramatized this by drawing what a man would look like if he could see as well as a normal man can see, but using compound eyes (Figure 5.22). The honeycomb pattern on the drawing is impressionistic, by the way. Each facet drawn actually stands for 10,000 ommatidia. The reason the man's compound eyes are only about one metre across instead of twenty-four is that Kirschfeld made allowance for the fact that we humans see very precisely only in the centre of our retina. He took an average of our precise central vision and our much less precise vision towards the edges of our retina, and came up with the one metre eye shown. Whether one metre across or twenty-four, a compound eye this large is impractical. The moral is, if you want to see precise, detailed images of the world, use a simple camera eye with a

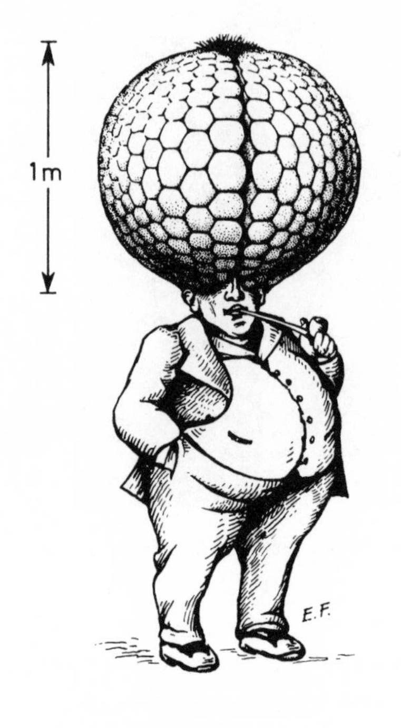

Figure 5.22 Kuno Kirschfeld's picture of how a man with compound eyes would look if he wanted to see as well as a normal human.

single, good lens, not a compound eye. Dan Nilsson even remarks of compound eyes that 'It is only a small exaggeration to say that evolution seems to be fighting a desperate battle to improve a basically disastrous design.'

Why, then, don't insects and crustaceans abandon the compound eye and evolve camera eyes instead? It may be one of those cases of becoming trapped the wrong side of a valley on the massif of Mount Improbable. To change a compound eye into a camera eye, there has to be a continuous series of workable intermediates: you cannot travel down into a valley as a prelude to mounting a higher peak. So, what is the problem about intermediates between a compound eye and a camera eye?

At least one outstanding difficulty comes to mind. A camera eye forms an upside-down image. A compound eye's image is the right way up. Finding an intermediate between those two is a tough proposition, to put it mildly. A possible intermediate is no image at all. There are some animals, living in the deep sea or otherwise in near total darkness, who have so few photons to play with that they give up on images altogether. All that they can hope for is to know whether there is light at all. An animal such as this could lose its image-processing nervous apparatus altogether and hence be in a position to make a fresh start up a completely different slope of the mountain. It could therefore constitute an intermediate on the path from a compound eye to a camera eye.

Some deep-sea crustaceans have large compound eyes but no lenses or optical apparatus at all. Their ommatidia have lost their tubes and their photocells are exposed right at the outer surface of where they will pick up what few photons there are, regardless of direction. From there it would seem but a small step to the remarkable eye of Figure 5.23. It belongs to a crustacean, called *Ampelisca*, which doesn't live particularly deep—perhaps it is on the way back up again from deep-sea ancestors. *Ampelisca*'s eye works as a camera eye, with a single lens forming an upside-down image on a retina. But the retina is clearly derived from a compound eye and consists of the remains of a bank of ommatidia. A small step, maybe, but only if, during the interregnum of near total blindness, the brain had enough evolutionary time to 'forget' all about processing right-way-up images.

That is an example of evolution from compound eye to camera eye (yet another example, by the way, of the ease with which eyes seem to evolve independently all around the animal kingdom). But

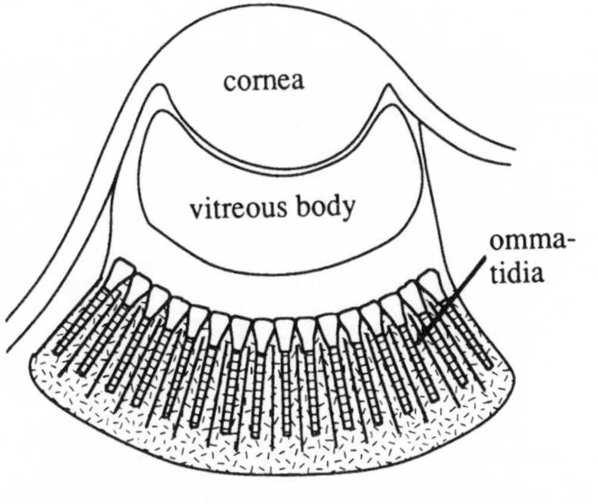

Figure 5.23 A camera eye with a compound eye in its ancestral history. The remarkable eye of *Ampelisca*.

how did the compound eye evolve in the first place? What do we find on the lower slopes of this particular peak of Mount Improbable?

Once again we may be helped by looking around the modern animal kingdom. Outside the arthropods (insects, crustaceans and their kin), compound eyes are found only in some Polychaet worms (ragworms and tubeworms) and in some bivalve molluscs (again, presumably independently evolved). The worms and molluscs are helpful to us as evolutionary historians because they also include among their number some primitive eyes which look like plausible intermediates strung out along the lower slopes of Mount Improbable leading to a compound-eye peak. The eyes in Figure 5.24 come from two different worm species. Once again, these are not ancestors, they are modern species and they are probably not even descended from the true intermediates. But they could easily be giving us a glimpse of what the evolutionary progression might have been like, from a loose clustering of photocells on the left to a proper compound eye on the right. This slope is surely just as gentle as the one we strolled up to reach the ordinary camera eye.

Ommatidia, as we have so far discussed them, depend for their effectiveness on being isolated from their neighbours. The gunsight that is looking at the dolphin's tailtip must not pick up rays from other parts of the dolphin, or we should be back with our original

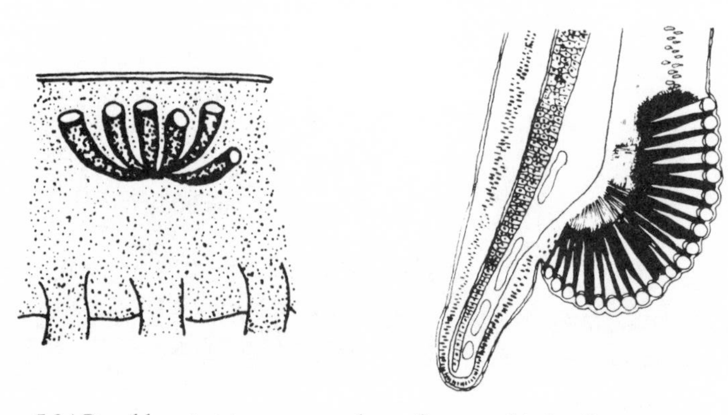

Figure 5.24 Possibly primitive compound eyes from two kinds of worms.

problem of millions of dolphin images. Most ommatidia achieve isolation by having a sheath of dark pigment around the tube. But there are times when this has bad side effects. Some sea creatures rely on transparency for their camouflage. They live in sea water and they look like sea water. The essence of their camouflage, then, is to avoid stopping photons. Yet the whole point of dark screens around ommatidia is to stop photons. How to escape from this cruel contradiction?

There are some deep-sea crustaceans who have come up with an ingenious partial solution (Figure 5.25). They don't have screening pigment, and their ommatidia are not tubes in the ordinary sense. Instead, they are transparent light guides, working just like man-made fibre-optic systems. Each light guide swells, at its front end, into a tiny lens, of varying refractive index like a fish eye. Lens and all, the light guide as a whole concentrates a large amount of light on to the

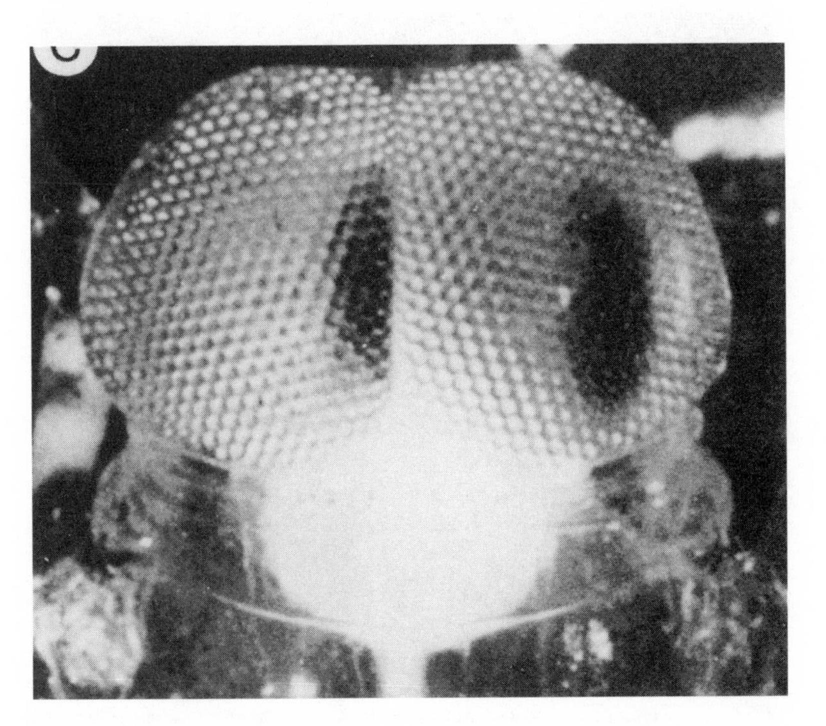

Figure 5.25 Eye of a deep-sea crustacean with fibre-optic light guides.

photocells at its base. But this includes only light coming from straight in the line of the gunsight. Beams coming sideways at a tube, instead of being shielded by a pigment screen, are reflected back and don't enter the light guide.

Not all compound eyes even try to isolate their private supply of light. It is only eyes of the apposition type that do. There are at least three different kinds of 'superposition' compound eye which do something more subtle. Far from trapping rays in tubes or fibre-optic light guides, they allow rays that pass through the lens of one ommatidium to be picked up by a neighbouring ommatidium's photocells. There is an empty, transparent zone, shared by all ommatidia. The lenses of all ommatidia conspire together to form a single image on a shared retina which is jointly put together from the light-sensitive cells of all the ommatidia. Figure 5.26 is Michael Land's picture of

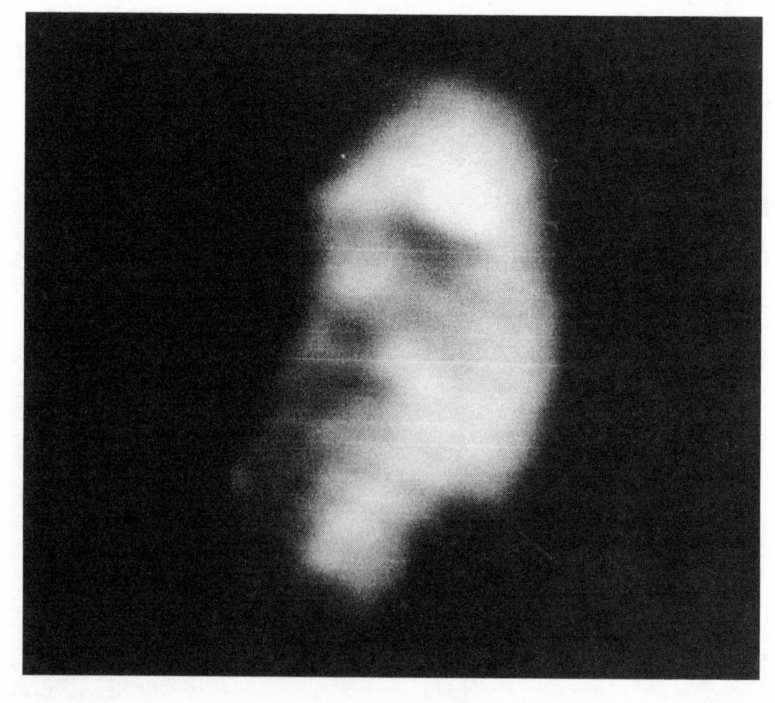

Figure 5.26 Charles Darwin's portrait, photographed by Michael Land through a firefly's compound lens.

Charles Darwin, seen through the compound lens of a firefly's superposition compound eye.

The image in a superposition compound eye, like that of apposition compound eyes but unlike that of camera eyes or that of Figure 5.23's *Ampelisca,* is the right way up. This is what you'd expect, assuming that superposition eyes are derived from apposition ancestors. It makes historical sense, and it must have made for an effortless transition as far as the brain was concerned. But it is still a very remarkable fact. For consider the physical problems of constructing a single right-way-up image in this way. Each individual ommatidium in an apposition eye has a normal lens in front of it which, if it makes an image at all, makes an upside-down one. To convert an apposition eye into a superposition one, therefore, the rays, as they pass through each lens, have somehow to be turned the right way up. Not only this, all the separate images from the different lenses have to be carefully superimposed to give one shared image. The advantage of doing this is that the shared image is much brighter. But the physical difficulties of turning the rays round are formidable. Amazingly, not only has the problem been solved in evolution, it has been solved in at least three independent ways: using fancy lenses, using fancy mirrors and using fancy neural circuitry. The details are so intricate that to spell them out would unbalance this already quite complicated chapter, and I'll deal with them only briefly.

A single lens turns the image upside-down. By the same token another lens, a suitable distance behind it, would turn it the right way up again. The combination is used in an instrument called a Keplerian telescope. The equivalent effect can be achieved in a single complex lens, using fancy gradations of refractive index. As we have seen, living lenses, unlike man-made ones, are good at achieving gradations of refractive index. This method of simulating the effect of a Keplerian telescope is used by mayflies, lacewings, beetles, moths, caddises and members of five different groups of crustaceans. The distance of their cousinship suggests that at least several of these groups evolved the same Keplerian trick independently of one another. An equivalent trick is pulled by three groups of crustaceans, using mirrors. Two of these three groups also contain members that do the lens trick. Indeed, if you look at which animal groups have adopted which of the several different kinds of compound eye, you notice a fascinating thing. The different solutions to problems pop up here, there and everywhere, suggesting, yet again, that they evolve rapidly and at the drop of a hat.

'Neural superposition' or 'wired-up superposition' has evolved in the large and important group of two-winged insects, the flies (fireflies, by the way, such as those mentioned in Figure 5.26, are not flies at all, but beetles). A similar system occurs in water boatmen, where it seems to have evolved—yet again—independently. Neural superposition is fiendishly cunning. In a way it shouldn't be called superposition at all, because the ommatidia are isolated tubes just as in apposition eyes. But they achieve a superposition-like effect by ingenious wiring of nerve cells behind the ommatidia. Here's how. You'll remember that the 'retina' of a single ommatidium is made up of about half a dozen photocells. In ordinary apposition eyes, the firing of all six photocells is simply added together, which is why I put 'retina' in quotation marks: all photons that shoot down the tube are counted, regardless of which photocell they hit. The only point of having several photocells is to increase the total sensitivity to light. This is why it doesn't matter that the tiny image at the bottom of an apposition ommatidium is technically upside-down.

But in the eye of a fly the outputs of the six cells are not pooled *with each other*. Instead, each one is pooled with the outputs of *particular* cells from *neighbouring* ommatidia (Figure 5.27). In the interests of clarity, the scale is all wrong in this diagram. For the same reason, the arrows don't represent rays (which would be bent by the lenses) but mappings from points on the dolphin to points in the bottoms of tubes. Now, see the shattering ingenuity of this scheme. The essential idea is that those photocells that are looking at the head of the dolphin in one ommatidium are ganged together with those photocells that are looking at the head of the dolphin in neighbouring ommatidia. Those photocells that are looking at the tail of the dolphin in one ommatidium are wired up together with those photocells that are looking at the tail of the dolphin in neighbouring ommatidia. And so on. The result is that each bit of the dolphin is being signalled by a

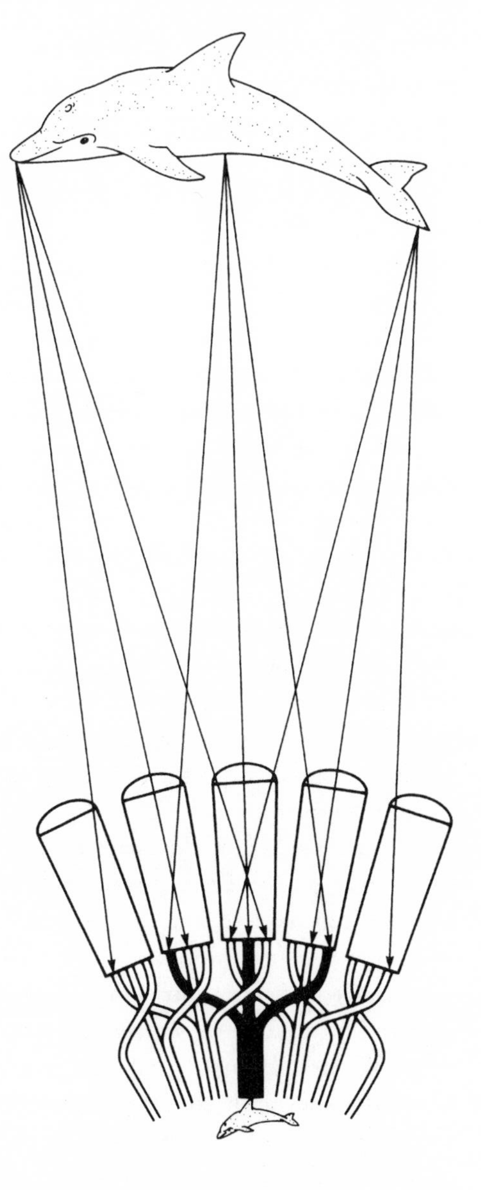

Figure 5.27 The ingenious principle of the 'wired-up superposition' compound eye.

larger number of photons than there would be in an ordinary apposition eye with a simple tube arrangement. It is a kind of computational, rather than an optical, solution to our old problem of how to augment the number of photons arriving from any one point on our dolphin.

You can see why this is called superposition, even though it strictly isn't. In true superposition, using fancy lenses or mirrors, light coming through neighbouring facets is superimposed so that photons from the dolphin's head end up in the same place as other photons from the dolphin's head; photons from the tail end up in the same place as other photons from the tail. In neural superposition, the photons still end up in different places, as they would in an apposition eye. But the signal *from* those photons ends up in the same place, due to the artful plaiting of the wires leading to the brain.

Nilsson's estimate for the rate of evolution of a camera eye was, you will recall, that it was by geological standards more or less instantaneous. You'd be lucky to find fossils that recorded the transitional stages. Exact estimates have not been done for compound eyes or any of the other designs of eye, but I doubt if they'd be significantly slower. One doesn't ordinarily expect to be able to see the details of eyes in fossils, because they are too soft to fossilize. Compound eyes are an exception because much of their detail is betrayed in the elegant array of more or less horny facets on the outer surface. Figure 5.28 shows a trilobite eye, from the Devonian era, nearly 400 million years ago. It looks just as advanced as a modern compound eye. This is what we should expect if the time it takes to evolve an eye is negligible by geological standards.

A central message of this chapter is that eyes evolve easily and fast, at the drop of a hat. I began by quoting the conclusion of one authority that eyes have evolved independently at least forty times in different parts of the animal kingdom. On the face of it, this message might seem challenged by an intriguing set of experimental results, recently reported by a group of workers in Switzerland associated with Professor Walter Gehring. I shall briefly explain what they found, and why it does not really challenge the conclusion of this chapter. Before I begin, I need to apologize for a maddeningly silly convention adopted by geneticists over the naming of genes. The gene called *eyeless* in the fruitfly *Drosophila* actually makes eyes! (Wonderful, isn't it?) The reason for this wantonly confusing piece of terminological contrariness is actually quite simple, and even rather interesting. We recognize what a gene does by noticing what happens when it goes

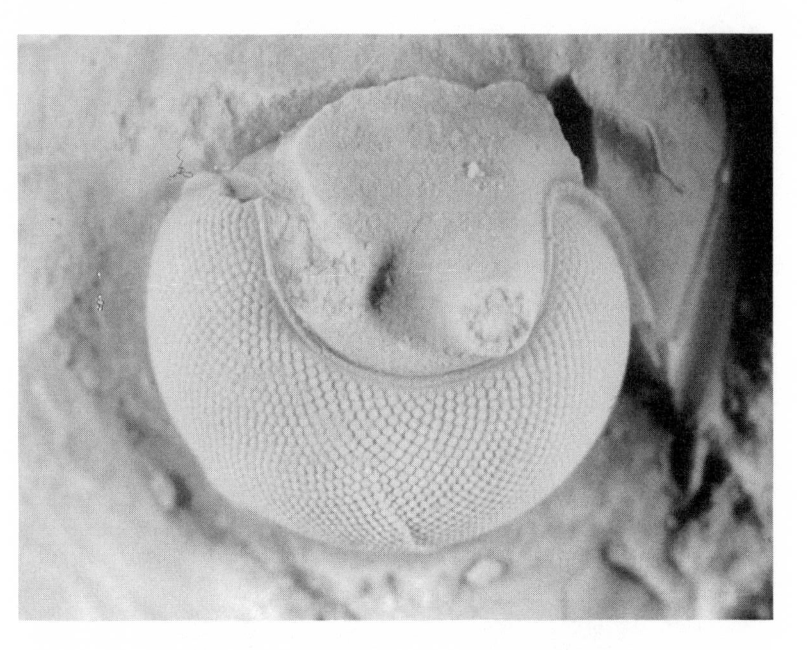

Figure 5.28 Compound eyes were already very advanced 400 million years ago: fossilized trilobite eye.

wrong. There is a gene which, when it goes wrong (mutates), causes flies to have no eyes. The position on the chromosome of this gene is therefore named the *eyeless* locus ('locus' is the Latin for 'place' and it is used by geneticists to mean a slot on a chromosome where alternative forms of a gene sit). But usually when we speak of the locus named *eyeless* we are actually talking about the normal, undamaged form of the gene at that locus. Hence the paradox that the *eyeless* gene makes eyes. It is like calling a loudspeaker a 'silence device' because you have discovered that, when you take the loudspeaker out of a radio, the radio is silent. I shall have none of it. I am tempted to rename the gene *eyemaker*, but this would be confusing too. I shall certainly not call it *eyeless* and shall adopt the recognized abbreviation *ey*.

Now, it is a general fact that although all of an animal's genes are present in all its cells, only a minority of those genes are actually turned on or 'expressed' in any given part of the body. This is why livers are different from kidneys, even though both contain the same complete set of genes. In the adult *Drosophila*, *ey* usually expresses itself only in the head, which is why the eyes develop there. George Halder, Patrick Callaerts and Walter Gehring discovered an experimental manipulation that led to *ey*'s being expressed in other parts of the body. By doctoring *Drosophila* larvae in cunning ways, they succeeded in making *ey* express itself in the antennae, the wings and the legs. Amazingly, the treated adult flies grew up with fully formed compound eyes on their wings, legs, antennae and elsewhere (Figure 5.29). Though slightly smaller than ordinary eyes, these 'ec-

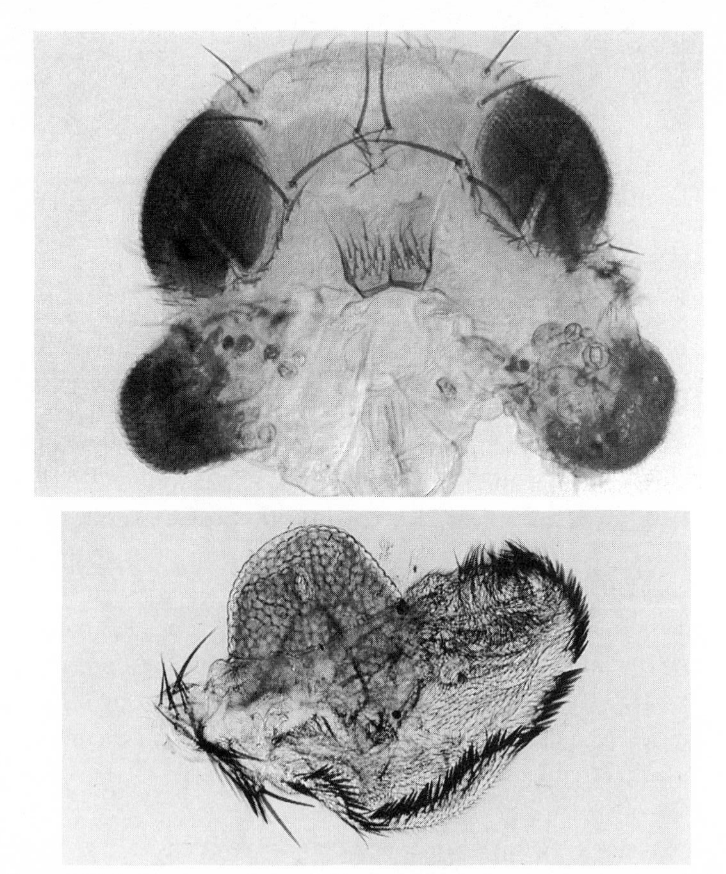

Figure 5.29 Induced ectopic eyes in Drosophilia; the bottom one has been induced by a mouse gene.

topic' eyes are proper compound eyes with plenty of properly formed ommatidia. They even work. At least, we don't know that the flies actually see anything through them, but electrical recording from the nerves at the base of the ommatidia shows that they are sensitive to light.

That is remarkable fact number one. Fact number two is even more remarkable. There is a gene in mice called *small eye* and one in humans called *aniridia*. These, too, are named using the geneticists' negative convention: mutational damage to these genes causes reduction or absence of eyes or parts of eyes. Rebecca Quiring and Uwe Waldorf, working in the same Swiss laboratory, found that

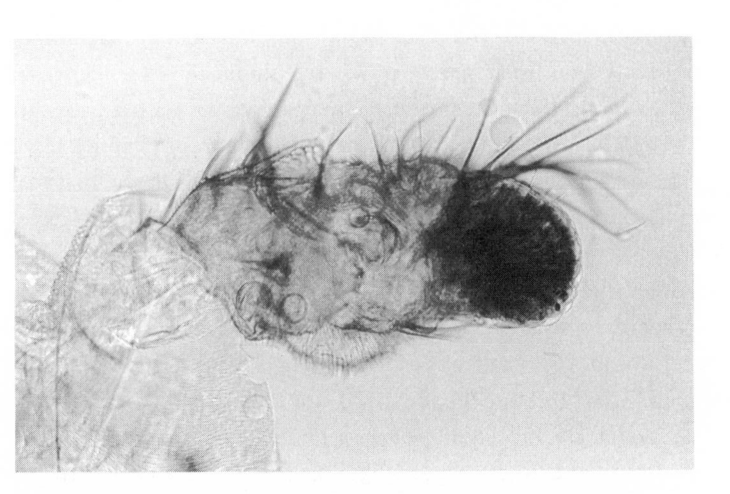

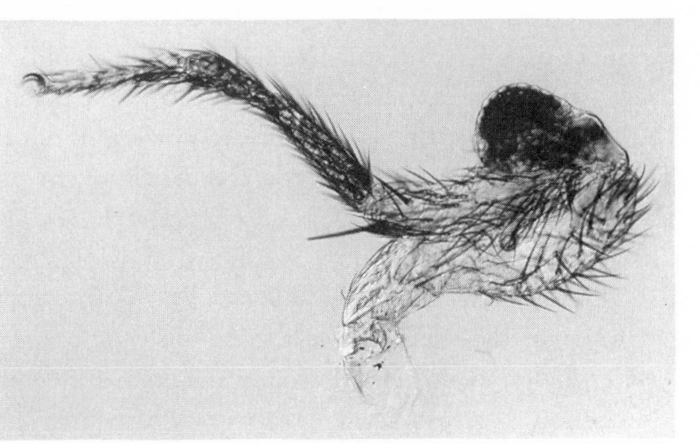

these particular mammal genes are almost identical, in their DNA sequences, to the *ey* gene in *Drosophila.* This means that the same gene has come down from remote ancestors to modern animals as distant from each other as mammals and insects. Moreover, in both these major branches of the animal kingdom the gene seems to have a lot to do with eyes. Remarkable fact number three is almost too startling. Halder, Callaerts and Gehring succeeded in introducing the mouse gene into *Drosophila* embryos. *Mirabile dictu,* the mouse gene induced ectopic eyes in *Drosophila.* Figure 5.29 (bottom) shows a small compound eye induced on the leg of a fruitfly by the mouse equivalent of *ey.* Notice, by the way, that it is an insect compound eye that has been induced, not a mouse eye. The mouse gene has simply switched on the eyemaking developmental machinery of *Drosophila.* Genes with pretty much the same DNA sequence as *ey* have been found also in molluscs, marine worms called nemertines, and sea-squirts. *Ey* may very well be universal among animals, and it may turn out to be a general rule that a version of the gene taken from a donor in one part of the animal kingdom can induce eyes to develop in recipients in an exceedingly remote part of the animal kingdom.

What does this spectacular series of experiments mean for our conclusion in this chapter? Were we wrong to think that eyes have developed forty times independently? I don't think so. At least the spirit of the statement that eyes evolve easily and at the drop of a hat remains unscathed. These experiments probably do mean that the common ancestor of *Drosophila,* mice, humans, sea-squirts and so on had eyes. The remote common ancestor had vision of some kind, and its eyes, whatever form they may have taken, probably developed under the influence of a sequence of DNA similar to modern *ey.* But the actual form of the different kinds of eye, the details of retinas and lenses or mirrors, the choice of compound versus simple, and if compound the choice among apposition or various kinds of superposition, all these evolve independently and rapidly. We know this by looking at the sporadic—almost capricious—distribution of these various devices and systems, dotted around the animal kingdom. In brief, animals often have an eye that resembles their remoter cousins more than it resembles their closer cousins. The conclusion remains unshaken by the demonstration that the common ancestor of all these animals probably had eyes of some kind, and that the embryonic development of all eyes seems to have enough in common to be inducible by the same DNA sequence.

After Michael Land had kindly read and criticized the first draft of this chapter, I invited him to attempt a visual representation of the eye region of Mount Improbable and Figure 5.30 shows what he drew. It is in the nature of metaphors that they are good for some purposes but not others and we must be prepared to modify them, or even drop them altogether, when necessary. This is not the first occasion when the reader will have noticed that Mount Improbable, for all that it has a singular name like the Jungfrau, is actually a more complicated, multiple-peaked affair.

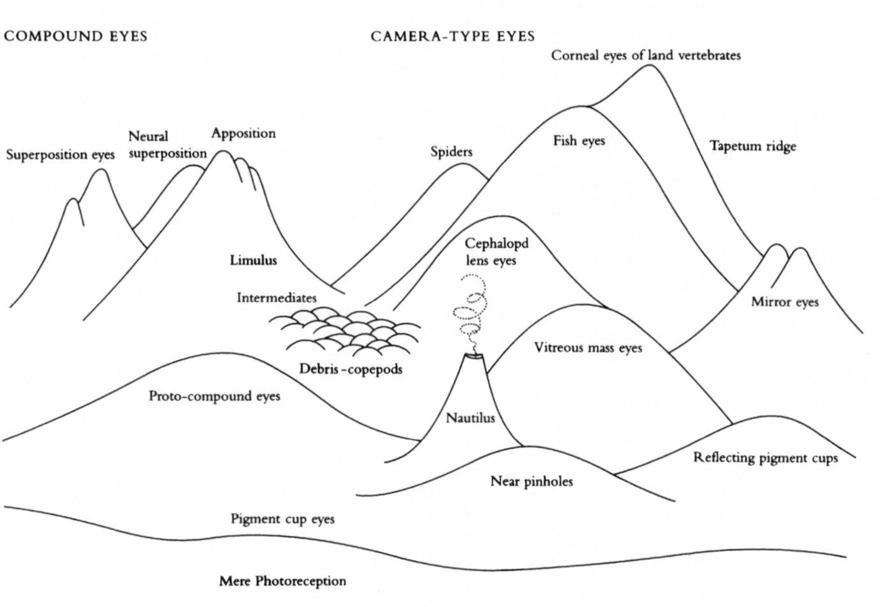

Figure 5.30 The eye region of the Mount Improbable range: Michael Land's landscape of eye evolution.

That other great authority on animal eyes, Dan Nilsson, who also read the chapter in draft, summed up the central message by calling my attention to what may be the most bizarre example of the *ad hoc* and opportunistic evolution of an eye. Three times independently, in three different groups of fish, the so-called 'four-eyed' condition has evolved. Probably the most remarkable of the four-eyed fish is *Bathylychnops exilis* (Figure 5.31). It has a typical fish eye looking outwards in the usual direction. But a secondary eye has evolved in addition, lodged in the wall of the main eye and looking straight downwards. What it looks at, who knows? Perhaps *Bathylychnops* suffers from a terrible predator with the habit of approaching from below. From our point of view the interesting thing is this. The embryological development of the secondary eye is completely different from that of the main eye, although we may surmise that its development may turn out to be induced in nature by a version of the *ey* gene. In particular, as Dr Nilsson put it in his letter to me, 'This species has re-invented the lens despite the fact it already had one. It serves as a good support for the view that lenses are not difficult to evolve.'

Nothing is as difficult to evolve as we humans imagine it to be. Darwin gave too much when he bent over backwards to concede the difficulty of evolving an eye. And his wife took too much when she underlined her scepticism in the margin. Darwin knew what he was

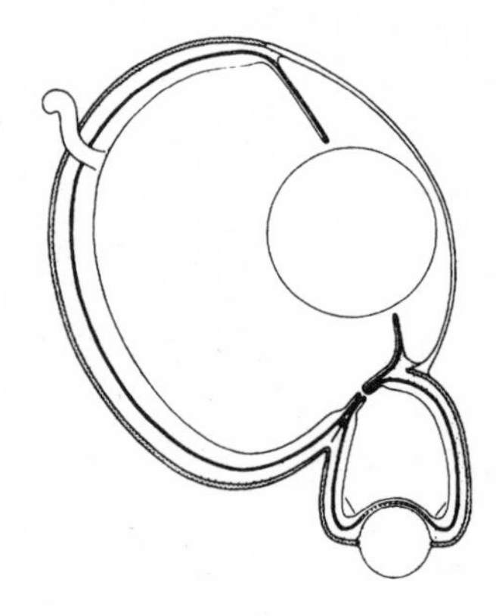

Figure 5.31 A remarkable double eye, that of the fish *Bathylychnops exilis*.

doing. Creationists love the quotation that I gave at the beginning of this chapter, but they never complete it. After making his rhetorical concession, Darwin went on:

> When it was first said that the sun stood still and the world turned round, the common sense of mankind declared the doctrine false; but the old saying of *Vox populi, vox Dei,* as every philosopher knows, cannot be trusted in science. Reason tells me, that if numerous gradations from an imperfect and simple eye to one perfect and complex, each grade being useful to its possessor, can be shown to exist, as is certainly the case; if further, the eye ever slightly varies, and the variations be inherited, as is likewise certainly the case; and if such variations should ever be useful to any animal under changing conditions of life, then the difficulty of believing that a perfect and complex eye could be formed by natural selection, though insuperable by our imagination, cannot be considered real.

CHAPTER 6

THE MUSEUM OF ALL SHELLS

NATURAL SELECTION IS THE PRESSURE THAT DRIVES evolution up the slopes of Mount Improbable. Pressure really is rather a good metaphor. We speak of 'selection pressure', and you can almost feel it pushing a species to evolve, shoving it up the gradients of the mountain. Predators, we say, provided the selection pressure that drove antelopes to evolve their fast running legs. Even as we speak, though, we remember what this really means: genes for short legs are more likely to end up in predators' bellies and therefore the world becomes less full of them. 'Pressure' from choosy females drove the evolution of male pheasants' sumptuous feathers. What this means is that a gene for a beautiful feather is especially likely to find itself riding a sperm into a female's body. But we think of it as a 'pressure' driving males towards greater beauty. No doubt predators provided a selection pressure in the opposite direction, towards duller plumage, since bright males would presumably attract predator, as well as female, eyes. Without the pressure from predators the cocks would be even brighter and more extravagant under pressure from females. Selection pressures, then, can push in opposite directions, or in the same direction or even (mathematicians can find ways of visualizing this) at any other 'angle' relative to one another. Selection pressures, moreover, can be 'strong' or 'weak', and the ordinary-language meanings of these words fit well. The particular path up

Mount Improbable that a lineage takes will be influenced by lots of different selection pressures, pushing and tugging in different directions and with different strengths, sometimes cooperating with each other, sometimes opposing.

But pressure isn't the end of the story. The path chosen up Mount Improbable will depend, too, on the shapes of the slopes. There are selection pressures, pushing and tugging in an assortment of directions and strengths, but there are also lines of least resistance and insurmountable precipices. A selection pressure may push for all it's worth in a particular direction, but if that direction is blocked by an impassable cliff it will come to nothing. Natural selection has to have alternatives to choose among. Selection pressures, however strong, can't do anything without genetic variation. To say that predators provide a selection pressure in favour of fast-running antelopes is just to say that predators eat the slowest antelopes. But if there is nothing to choose between the genes of fast and slow antelopes—that is, if differences in running speed are purely environmentally determined—no evolutionary business will result. In the direction of improved speed, Mount Improbable might present no slope to climb.

Now we come to a piece of genuine uncertainty and a spectrum of opinion among biologists. At one extreme are those who feel that we can take genetic variation more or less for granted. If the selection pressure exists, they feel, there will always be enough genetic variation to accommodate it. The trajectory of a lineage in evolutionary space will be, in practice, determined by the tussle among selection pressures alone. At the other extreme are those who feel that available genetic variation is the important consideration determining the direction of evolution. Some even go so far as to assign natural selection a minor, subsidiary role. To take our two biologists to the point of caricature, we might imagine them disagreeing on why pigs don't have wings. The extreme selectionist says that pigs don't have wings because it would not be an advantage for them to have wings. The extreme anti-selectionist says that pigs might benefit from having wings, but they can't have them because there never were mutant wing stubs for natural selection to work upon.

The controversy is more sophisticated than that, and Mount Improbable, even in its multiple-peaked version, isn't a powerful enough metaphor to explore it. We need a new metaphor, using the kind of imagination that mathematicians enjoy although we shan't use explicit mathematical symbols. It will make more demands on us than Mount Improbable, but it is worth it. In *The Blind Watchmaker* I made brief excursions into what I variously called 'genetic space', 'biomorph land' and 'Making Tracks Through Animal Space'. More recently the philosopher Daniel Dennett has penetrated further into this undiscovered country which, by poetic allusion to Borges's Library of Babel, he calls the Library of Mendel. My version in this chapter is a gigantic museum of the zoological imagination.

Imagine a museum with galleries stretching towards the horizon in every direction, and as far as the eye can see upwards and downwards as well. Preserved in the museum is every kind of animal form that has ever existed, and every kind that could be imagined. Each animal is housed next door to those that it most resembles. Each dimension in the museum—that is, each direction along which a gallery extends—corresponds to one dimension in which the animals vary. For example, as you walk north along a particular gallery you notice a progressive lengthening of the horns of the specimens in the cabinets. Turn round and walk south and the horns shorten. Turn and walk east and the horns stay the same but something else changes, say the teeth get sharper. Walk west and the teeth grow blunter. Since horn length and tooth sharpness are only two out of thousands of ways in which animals can vary, the galleries must criss-cross one another in many-dimensional space, not just the ordinary three-dimensional space that we, with our limited minds, are capable of visualizing. This is what I meant when I said that we had to learn to think like a mathematician.

What would it mean to think in four dimensions? Suppose we are dealing with antelopes and we measure four variables: horn length, tooth sharpness, intestine length and coat hairiness. If we ignore one of the dimensions, say coat hairiness, we could place each of our antelopes in its rightful place in a three-dimensional graph—a cube—of the remaining variables, horn length, tooth sharpness and intestine length. Now how do we bring in the fourth dimension, coat hairiness? We do the whole cube exercise separately for all short-haired antelopes, then we produce another cube for all slightly longer-haired antelopes and so on. A given antelope will be placed, first in whichever cube pertains to its hair length and then, within that cube, to its rightful position determined by its horns, teeth and intestines. Coat hairiness is the fourth dimension. In principle you can go on constructing families of cubes, and cubes of cubes, and cubes of cubes of cubes until you have placed animals in the equivalent of many-dimensional space.

To get some idea of what we are supposed to be thinking of when we think of the Museum of All Possible Animals, this chapter will deal with a particular case which can, more or less, be confined to three dimensions. In the next chapter I shall return to the controversy with which this chapter began and try to make a constructive overture towards the other side (for I am a known partisan). This chapter's three-dimensional special case is that of snail shells and other coiled shells. The reason the galleries of shells can be confined to three dimensions is that most of the important variation among shells can be expressed as change in only three numbers. In what follows, I shall be following in the footsteps of David Raup, a distinguished palaeontologist from the University of Chicago. Raup, in turn, was inspired by the celebrated D'Arcy Wentworth Thompson, of the ancient and distinguished Scottish University of St Andrews, whose book, *On Growth and Form* (first published in 1919), has been a persistent, if not quite mainstream, influence on zoologists for most of the twentieth century. It is one of the minor tragedies of biology that D'Arcy Thompson died just before the computer age, for almost every page of his great book cries out for a computer. Raup wrote a program to generate shell form, and I have written a similar program to illustrate this chapter although—as might be expected—I incorporated it in a Blind Watchmaker-style artificial selection program.

The shells of snails and other molluscs, and also the shells of creatures called brachiopods which have no connection with molluscs but superficially resemble them, all grow in the same kind of way, which is different from the way we grow. We start small and grow all over (with some bits growing faster than others). You can't take a man and dissect out the bit of him that was him as a baby. With a mollusc shell you can

Figure 6.1 A section through the shell of a Nautilus. The animal itself lives in the largest, most recent, chamber.

do just that. A mollusc shell starts small and grows at the margins, so the innermost part of the adult coil is the baby shell. Each animal carries its own infant form around with it, as the narrowest part of its shell. The shell of *Nautilus* (already mentioned for its pinhole eye) is divided into air-filled flotation compartments, all except for the largest and most recently built compartment at the growing margin, in which the animal itself lives at any one time (Figure 6.1).

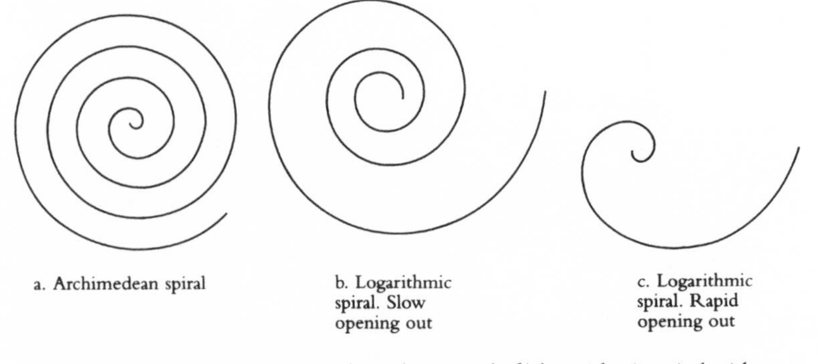

Figure 6.2 Kinds of spirals: *(a)* Archimedean spiral; *(b)* logarithmic spiral with slow rate of opening; *(c)* logarithmic spiral with rapid rate of opening.

Because of their method of expanding at the margin, shells all have the same general form. It is a solid version of the so-called logarithmic or equiangular spiral. The logarithmic spiral is different from the Archimedean spiral which is what a sailor produces when he coils a rope on the deck. No matter how many turns the rope takes, each successive turn is still the same width—one thickness of the rope. In a logarithmic spiral, by contrast, the spiral opens out as it propels itself away from the centre. Different spirals open out at different rates, but it is always a particular rate for any particular spiral. Figure 6.2 shows, in addition to an Archimedean coiled-rope spiral, two logarithmic spirals with different rates of opening out.

A shell grows, not as a line, but as a tube. The tube doesn't have to be circular in cross-section like a French horn but, just for the moment, we'll assume that it is. We'll also assume that the spiral drawn represents the outer margin of the tube. The diameter of the tube could happen to expand at just the right rate to keep the inner margin fitting snugly against the previous whorl of the spiral, as in Figure 6.3a. But it doesn't have to. If the tube's diameter expands more slowly than the outer margin of the spiral, a gap of increasing size will be left between successive whorls, as in Figure 6.3b. The more 'gappy' the shell, the more it seems suitable for a worm rather than for a snail.

Raup described the spirals of shells using three numbers, which he called *W*, *D* and *T*. I hope it will not be thought too quaint if I re-

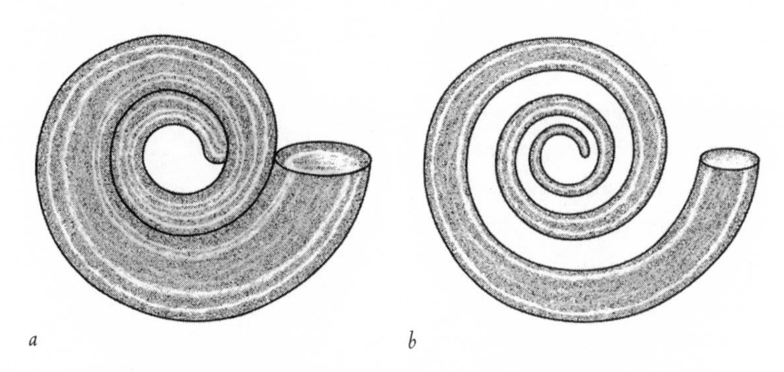

Figure 6.3 Two tubes with same spiral but different tube size: *(a)* tube big enough to fill the gap between successive whorls of the spiral; *(b)* tube narrow enough to leave thin air (or open water) between successive whorls of the spiral.

name them *flare, verm* and *spire.* It is easier to remember which is which than in the case of the mathematical letters. *Flare* is a measure of the expansion rate of the spiral. If the *flare* is 2, this means that, for every complete turn around the spiral, the spiral opens out to twice its previous size. This is true of Figure 6.2b. For every turn around Figure 6.2b, the width of the spiral doubles. Figure 6.2c, being a much more open shell, has a *flare* of 10. For each complete circuit around this spiral, the width would increase tenfold (although in practice the spiral comes to an end before it has time to complete a whole circuit). Something like a cockle, which opens out so rapidly that you don't even think of it as coiling, has a *flare* value up in the thousands.

When describing *flare* I was careful not to say that it measures the rate of increase of the diameter of the tube. This is where the second number, *verm,* comes in. We need *verm* because the tube does not have to fill, snugly, the space made available by the expanding spiral. The shell can be 'gappy', like the one in Figure 6.3b. *Verm* gets its name from 'vermiform' which means 'worm-shaped'. Figure 6.3a and Figure 6.3b have identical *flare* values (2) but Figure 6.3b has a higher *verm* score (0.7) than Figure 6.3a's 0.5. A *verm* of 0.7 means that the distance from the centre of the spiral to the inner margin of the tube is 70 per cent of the distance from the centre of the spiral to the outer margin of the tube. It doesn't matter which part of the tube you use to make the measurement, the *verm* score is the same (this doesn't logically have to be true but it seems often to be true of real shells and we shall assume it unless otherwise stated). You can easily see that a very high *verm* like 0.99 makes for a very thin, threadlike tube, because the inner margin of the tube is 99 per cent of the distance to the outer margin of the tube.

What *verm* value is needed to ensure snug fitting, as in Figure 6.3a? That depends on the *flare.* To be precise, the critical *verm* value for a snug fit is exactly the reciprocal of the *flare* value (that is, one divided by *flare*). The *flare* is 2 in both parts of Figure 6.3, so the critical *verm* for a snug fit is 0.5, and this is what Figure 6.3a has. Figure 6.3b has a *verm* which is higher than its 'snug critical' value, which is why the shell appears gappy. If we take a shell like Figure 6.2c with a *flare* value of 10, the snug critical *verm* score would be 0.1.

What if the *verm* value is smaller than the snug critical value? Could we imagine a tube so fat that it goes beyond snug fitting and actually encroaches inside the territory of the previous whorl—for example, a spiral like those in Figure 6.3 but with a *verm* value of, say, 0.4? There are two ways in which the clash can be resolved. One is simply to allow the tube to enclose earlier whorls of itself. *Nautilus* does this. It means that the shape of the cross-section of the available tube can no longer be a plain circle but has a 'bite' taken out of it. But this is no disaster because, as you'll remember, it was only an arbitrary decision to assume that the tube has a circular section in any case. Many molluscs live happily in a tube which is far from circular in section, and we shall come on to them. In some cases the best way to interpret the non-circular shape of the cross-section of the tube is as a means of accommodating previous whorls of the tube.

The other way to resolve the would-be encroachment of previous whorls of the tube is to move out of the plane. This brings us to the third of our shell signature numbers, *spire*. Think of the expanding spiral as moving sideways as it expands, making a conical shape like a top. The third shell signature number, *spire*, is the rate at which successive whorls of the spiral creep along the length of the cone. *Nautilus* happens to have a *spire* value of 0: all its successive windings are in one plane.

So, we have three shell signature numbers, *flare*, *verm* and *spire* (Figure 6.4). If we ignore one of these, say *spire*, we can plot a graph of the other two on a flat piece of paper. Every point on the graph has a unique combination of *flare* and *verm* values, and we can program the computer to draw, at that point, the shell that would be produced. Figure 6.5 shows twenty-five regularly spaced points on the graph. As you move from left to right across the graph, the computer shells become progressively more 'wormy' as *verm* increases. As you move from top to bottom and *flare* increases, the spirals become progressively more open until they don't look obviously spiral at all. In order to get a good spread as we go down, we make *flare* increase logarithmically. This means that each equal step down the page corresponds to *multiplying* by some number (in this case ten) rather than, as in a normal graph and as in the progression

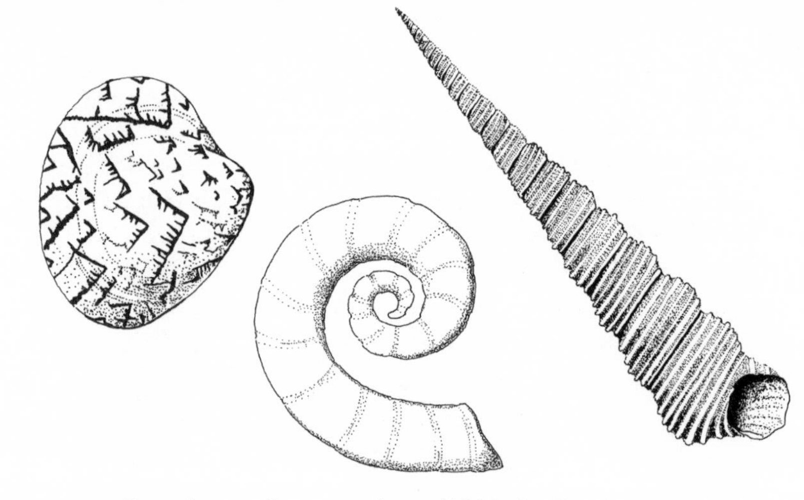

Figure 6.4 Shells to illustrate *flare, verm* and *spire*: *(a)* high *flare*: *Liconcha castrensis*, a bivalve mollusc; *(b)* high *verm*: *Spirula*; *(c)* high *spire*: *Turritella terebra.*

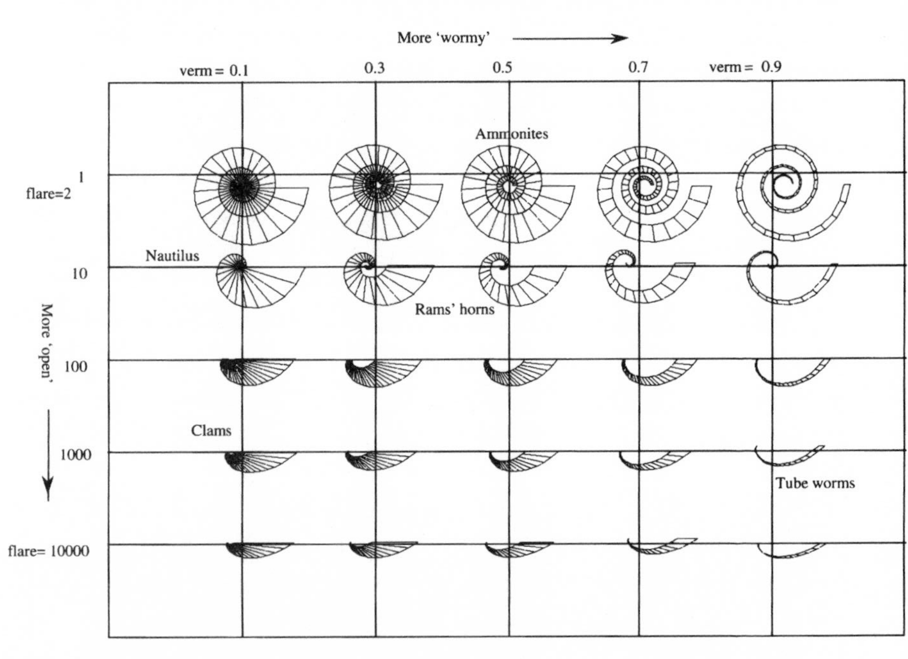

Figure 6.5 Table of computer-generated shells systematically varying *verm* and *flare*. Changes in the third dimension, *spire*, would not be visible in this view. The *flare* axis is logarithmic—equal steps down the page represent a tenfold multiplication of the value of *flare*. On the *verm* axis, equal steps across the page represent a fixed addition to the 'worminess' score. A few named real animals are written in approximately their right places on the chart.

of *verm*-scores across the page, *adding* a number with each step. This is necessary in order to accommodate shells like cockles and clams, at the bottom left of the picture (which have *flare* values up in the thousands where small changes don't make much difference) in the same graph as ammonites and snails (which typically have *flare* scores in low, single figures where small changes make a big difference). In various parts of the graph you can see shapes that resemble ammonites, *Nautilus*, clams, rams' horns and tubeworms, and I've written labels in approximately the right places.

My computer program can draw shells in two views. Figure 6.5 shows one view, emphasizing the shape of the spiral itself. Figure 6.6 shows the other view, 'X-ray' cross-sections, giving an impression of the solid shape of the shells. Figure 6.7 is an actual X-ray photograph of real shells to explain the nature of this view. The four shells of Figure 6.6 are computer shells chosen, like the real shells of Figure 6.4, to illustrate different values of *flare*, *verm* and *spire*.

Figure 6.8 is a graph, similar to Figure 6.5 except that the computer shells are shown in X-ray view, and the axes are *flare* and *spire* instead of *flare* and *verm*.

One could also, of course, plot *verm* versus *spire*, but I won't take the space to do this. Instead, I'll go straight to Raup's famous cube (Figure 6.9). Because three numbers suffice to define a shell (leaving aside the question of cross-sectional shape of the tube) you can place each shell in its own unique spot in a three-dimensional box. The Museum of Possible Shells, unlike, say, the Museum of Possible Pelvis Bones, is a simple tower block. One dimension corresponds to each of the three shell signature numbers. Stand in the Museum of Possible Shells and walk, say, north, which we'll designate the *verm* dimension. As you make your way along the gallery, the shells that you pass steadily become more 'wormy', while keeping everything else constant. If at any point you turn left and walk west, the shells that you pass steadily increase their *spire* value, becoming more cone-shaped, while keeping other things constant. Finally if, at any point, you stop moving either west/east or north/south and climb directly downwards instead—the *flare* dimension—you encounter shells with a steadily increasing rate of opening out. You can get from any shell

flare=2, verm=0, spire=3

flare=1.3, verm=0, spire=8.2

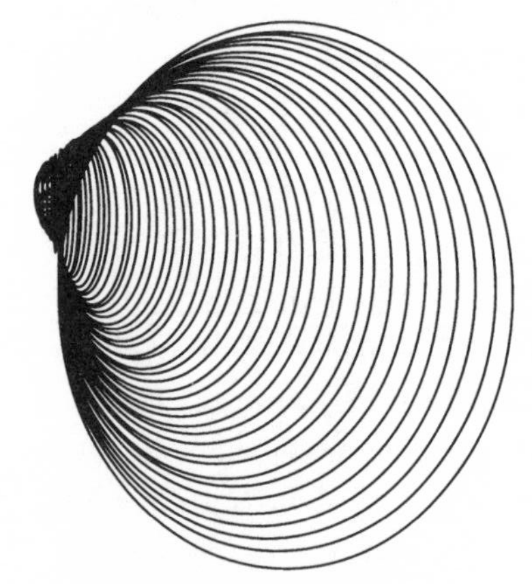

flare=1000, verm=0, spire=0.5

flare=2, verm=0.25, spire=1.5

Figure 6.6 Four computer shells in 'X-ray' view, to show different *flare, verm* and *spire* values.

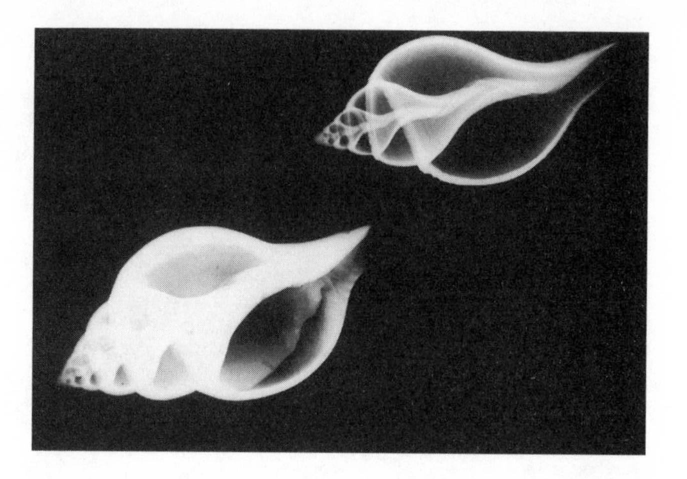

Figure 6.7 Photograph of a real shell in X-ray view.

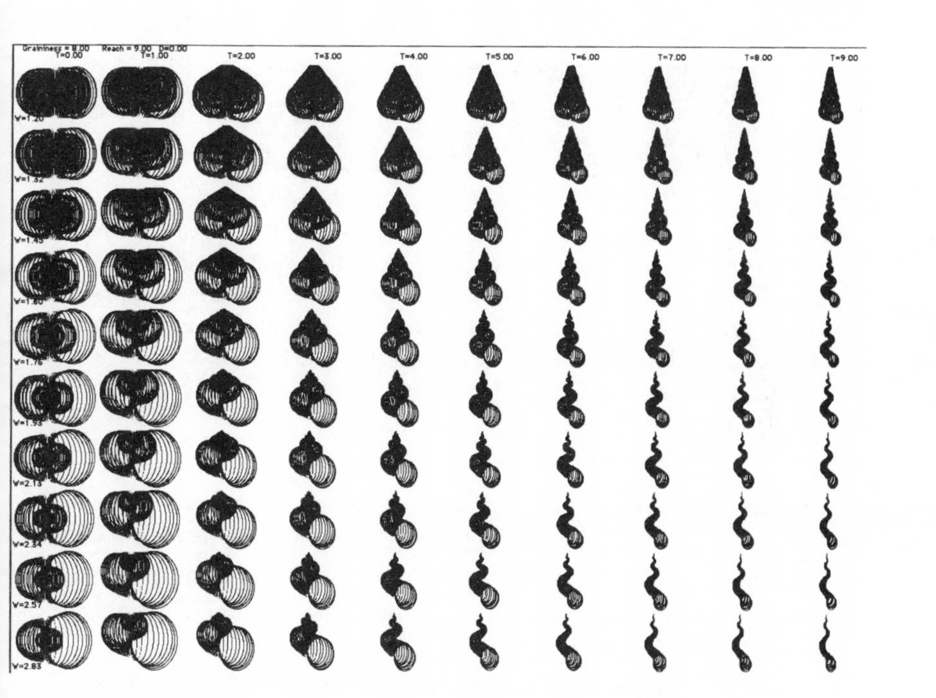

Figure 6.8 Graph of computer shells ('X-ray view') plotting *flare* (labelled W, down the page) versus *spire* (labelled T, across the page). As in Figure 6.5, the *flare* scale is logarithmic, but here *flare* is confined to low values—none of the shells opens out very far.

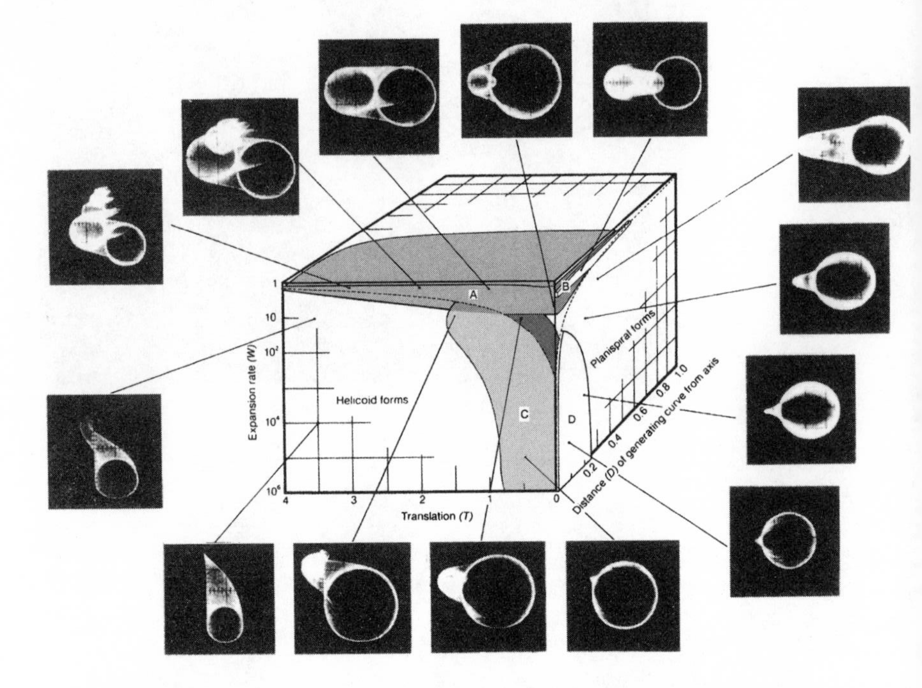

Figure 6.9 Raup's cube. David M. Raup drew a three-dimensional graph of *flare* (which he called W) down the page, against *spire* (which he called T) across the page from right to left, against *verm* (which he called D) backwards into the page. 'X-ray views' of computer-drawn shells are sampled at strategic points in the cube. Regions of the cube in which real-life shells can be found are shaded. The unshaded regions house theoretically conceivable shells that do not actually exist.

to any other by burrowing through the cube at the appropriate angle, and you'll pass a continuous series of intermediate shells on the way. Figures 6.5 and 6.8 can be thought of as two outer faces of Raup's cube. On two-dimensional paper we could print any slice, at any particular angle, through the cube.

Raup wrote the original computer program that is the inspiration for mine. In his published diagram, rather than attempt the impractical task of drawing all the shells in the cube, Raup sampled particular points. The pictures round the edge of Figure 6.9 represent the theoretical shells that you would find at the designated points of the space. Some of them look like actual shells that you might find on a beach. Others look like nothing on earth, but they still belong in the space of all computable shells. Raup shaded in on his picture those regions of the space where actual shells are to be found.

Ammonites, those once-ubiquitous *Nautilus*-relatives who seem to have come to the same sad end (whatever it was) as the dinosaurs, had coiled shells but, unlike snails, their coils were nearly always limited to one plane. Their *spire* value was zero. At least, this is true of typical ammonites. Pleasingly, however, a few of them, such as the Cretaceous genus *Turrilites* evolved a high *spire* value, thereby independently inventing the snail form. Such exceptional forms apart, the ammonites are housed along the eastern wall of the Museum of Shells (names like 'east' and 'south' are, of course, arbitrary labels for the diagram). The cabinets of typical ammonites don't occupy more than the southern half of the eastern wall, and only the top few storeys. Snails and their kind overlap with the Ammonite Corridor but they also spread far to the west (the *spire* dimension) and they penetrate a little further down towards the lower storeys of the tower block. But most of the lower storeys—where the *flare* rate is large and shells open out rapidly—belong to the two great groups of double-shelled creatures. Bivalve molluscs stretch a little to the west—they have a slight twist on them like snails but their tube opens out so fast that they don't look like snails. Brachiopods or 'lamp-shells', which, as we have seen, are not molluscs at all but superficially resemble bivalve molluscs, share with ammonites a 'coil' that is entirely in one plane. As with the molluscan bivalves, brachiopod tubes typically flare completely open before they have time to build up a 'coil' worthy of the name.

Any particular evolutionary history is a trajectory through the Museum of All Possible Shells and I have represented this by embedding my shell-drawing computer procedure in the larger, Blind Watchmaker artificial-selection program. I simply removed the tree-growing embryology from the Blind Watchmaker program and dropped a shell-growing embryology into its place instead. The combined program is called Blind Snailmaker. Mutation is equivalent to small movements in the museum—remember that all shells are surrounded by their most similar neighbours. In the program, the three shell signature numbers are each represented by one gene locus whose numerical value can vary. So we have three classes of mutation, small changes in *flare,* small changes in *verm* and small changes in *spire.* These mutational changes can be positive or negative, within limits. The *flare* gene has a minimum value of 1 (smaller values would indicate a shrinking rather than a growth process) and no fixed maximum value. The *verm* gene's value is a proportion, varying from 0 to just below 1 (a *verm* of 1 would indicate a tube so thin and wormy as to be non-existent). *Spire* has no limits: negative values trivially indicate an upside-down shell. Following the original Blind Watchmaker program, Blind Snailmaker presents a parent shell in the middle of the computer screen, surrounded by a litter of asexual offspring—its randomly mutated near neighbours in the Museum of All Shells. The human selector clicks the mouse to choose one of the shells for breeding. It glides to the parental position in the centre, and the screen fills up with a litter of its offspring. The process recycles as long as the selector has patience. Slowly, you feel yourself creeping through the Museum of All Possible Shells. Sometimes you are walking amidst familiar shells, of the kind that you could pick up on any beach. At other times you stray outside the bounds of reality, into mathematical spaces where no real shells have ever existed.

I earlier explained that, although the set of all possible shells can largely be described with only three numbers, this does include a simplifying assumption which is wrong: the assumption that the cross- sectional shape of the tube is always a circle. It seems to be generally true that, as the tube flares out, it remains the same shape, but it is by no means true that that shape is always a circle. It can be an oval and my computer model incorporates a fourth 'gene', called *shape,* whose value is the height of the oval tube divided by its width. A circle is the special case of a *shape* of 1. Incorporation of this gene adds surprisingly to the power of the model to represent real shells. But it still is not enough. Many real shells have a variety of more complicated cross-sectional shapes which are neither circular nor oval, and which don't lend themselves to simple mathematical description. Figure 6.10 shows a range

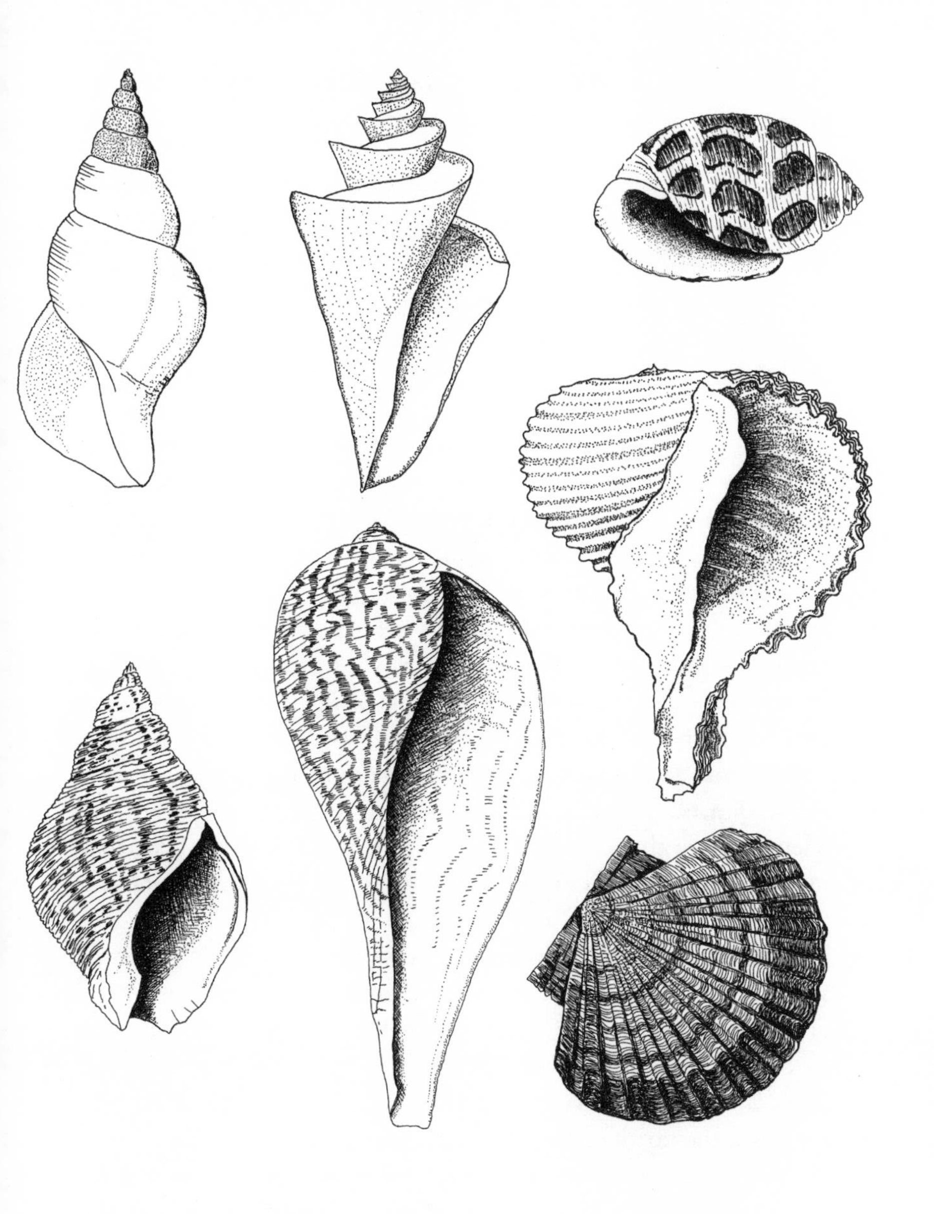

Figure 6.10 Real shells with a range of cross-sectional shapes: (clockwise from bottom left): speckled whelk, *Cominella adspersa*; left-handed Neptune, *Neptunia contraria*; Japanese wonder shell, *Thatcheria mirabilis*; the Eloise, *Acteon eloisae*; Rapa snail, *Rapa rapa*; great scallop, *Pecten maximus*; graceful fig shell, *Ficus gracilis*.

of shells which, as well as coming from different parts of the cubic museum, also have complicated, non-circular cross-sections of their basic tube.

My Blind Snailmaker program incorporates this extra variation by the rather crude expedient of providing a repertoire of pre-drawn cross-sectional outlines. Each of these outlines is then transformed (flattened vertically or horizontally) by the current (and mutable) value of the gene *shape*. The program then generates a tube of that transformed outline, sweeping it around and out just as if it were a circular tube. A better way to handle this problem—and one which I might attempt one day—would be to program the computer to simulate the actual growth process varying around the leading edge of the tube and thereby form ornate cross-sections. Nevertheless, for what it is worth, Figure 6.11 is a 'zoo' of computer shells produced by the existing program, through artificial selection using the human eye. They were bred for resemblance to known shells, some of them approximately similar to those of Figure 6.10, some of them similar to other shells that you might find on a beach or while diving.

The cross-sectional shape of the tube can be regarded as an additional dimension (or set of dimensions) in the Museum of All Shells. Setting that on one side and reverting to our simplifying assumption of a circular cross-section, one of the beauties of shells is that they are easy to fit into a Museum of All Possible Forms that we can actually draw in three dimensions. But this doesn't mean that all parts of the theoretical museum are tenanted in real life. In real life, as we have seen, most of the volume of the museum tower block is empty. Raup shaded the lived-in regions (Figure 6.9), and they constitute much less than half the volume of the cube. Stretching far to the north and west, gallery after gallery houses hypothetical shells that could exist according to the mathematical model but which actually have never been seen on this planet. Why not? And why, since we are asking such questions, are the shells that *have* really existed confined to this particular cuboidal building in the first place?

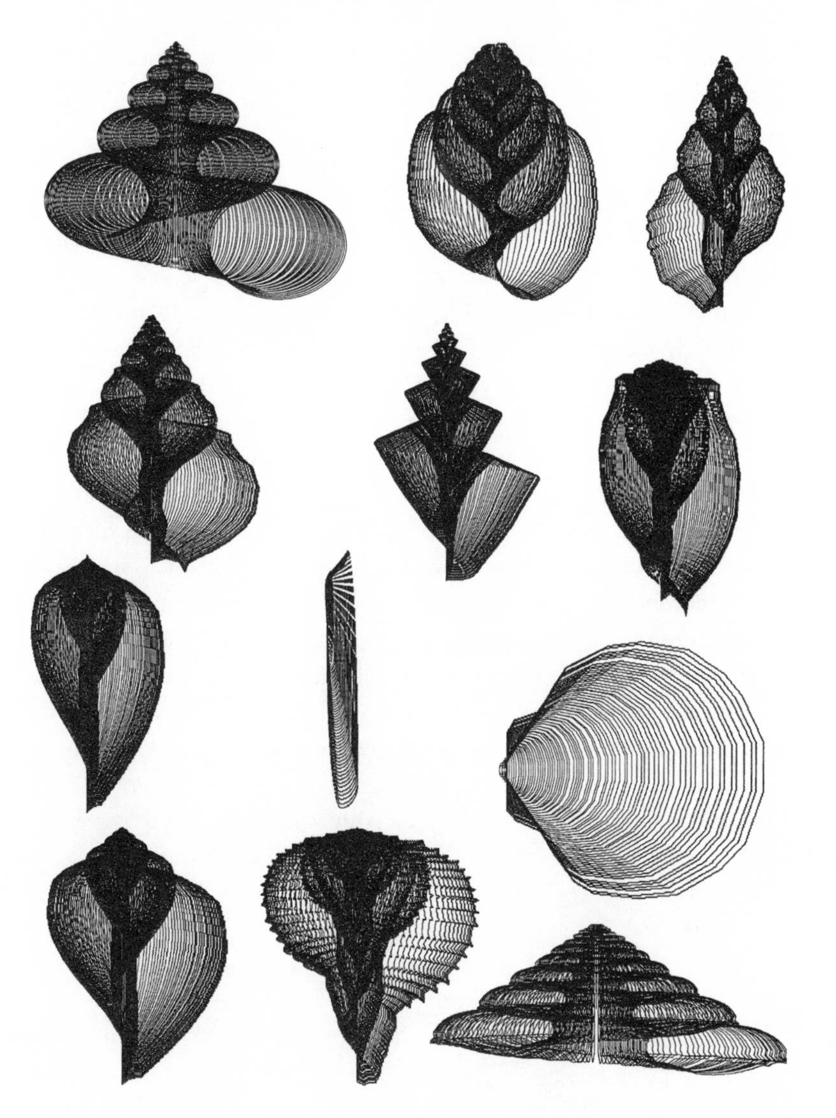

Figure 6.11 'Zoo' of computer shells of different cross-sectional shapes bred using the Blind Snailmaker program. They were bred by artificial selection, the eye choosing them for their resemblance to familiar real shells, including some members of the range depicted in Figure 6.10.

What might a shell look like if it did not fit into the mathematical tower block? Figure 6.12a shows a computer-generated snail that doesn't. Instead of having a fixed *spire* value, its *spire* value changes as it grows older. More recent, wider, parts of the shell grow with a lower *spire* value than developmentally earlier, narrower, parts of the shell. This is why the whole shell has such an 'unnaturally', and presumably vulnerably, pointed top. This snail is hypothetical. It exists in the computer only. The computer 'cone' shell in Figure 6.12b also has an unnaturally pointed top. It too was drawn by the Blind Snailmaker program, but with the *spire* value programmed to decrease, rather than remain constant, as development proceeded.

The shells in Figure 6.13 are real, and I suspect that they too have a *spire* gradient, meaning that they begin life with a high *spire* value and gradually decrease it as they grow older. According to Raup, there were some real ammonites that changed their shell signature numbers as they grew older. You could say that as they grow older these odd shells move from one part of the museum to another and that they still stay within the museum. But it is also true to say that, since the juvenile body is included as part of the adult one, there is no one cabinet in the museum where the whole body can be housed. People could disagree over whether the animals in Figure 6.13 should be regarded as truly confined to the three dimensions of the box. Geerat Vermeij, one of today's leading experts

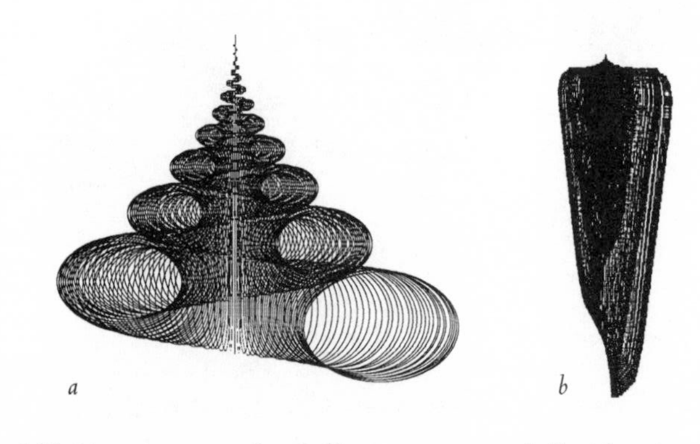

Figure 6.12 *(a)* computer snail and *(b)* computer cone-shell with pointy *spires* produced by a 'gradient' on the *spire* gene.

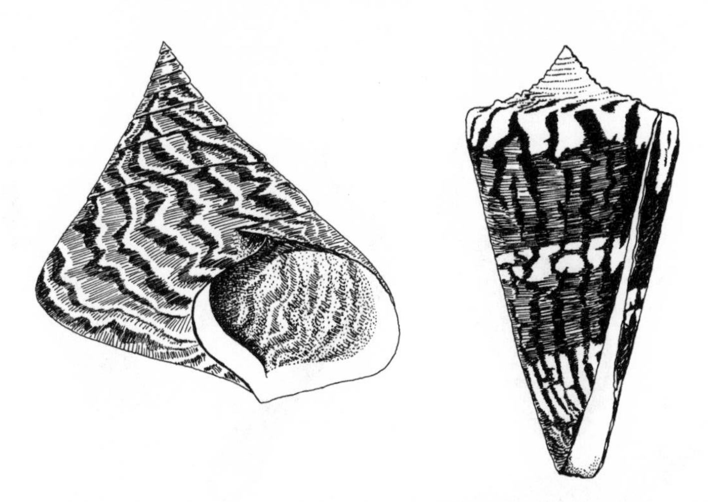

Figure 6.13 Real shells whose resemblance to the computer shells of the previous figure suggests that they too develop with *spire* gradients. Left: tiger maurea, *Maurea tigris*; right: general cone, *Conus generalis*.

on the zoology of shelled animals, believes that a tendency to change signature numbers as the animals grow older may be the norm rather than the exception. He believes, in other words, that most molluscs shift their position in the mathematical museum, at least a little bit, as they grow up.

Let us turn to the opposite question of why large areas of the museum are empty of real shells. Figure 6.14 shows a sample of computer-generated shells from deep in the 'no-go' regions of the museum. Some of them might look fine on the head of an antelope or bison, but as mollusc shells they have never seen the light of day. With the question of why there are no shells like this, we come right back to the controversy with which we began. Is evolution limited by lack of available variation, or is it that natural selection doesn't even 'want' to visit certain areas of the museum? Raup himself interpreted the empty areas—the unshaded zones of his cube—in selectionist terms. There is no selection pressure on shellfish to move into the areas represented by the gaps. Or, to put it another way, shells with those theoretically possible shapes would, in practice, have been bad shells in which to live: perhaps weak and easily crushed; or otherwise vulnerable, or uneconomical of shell material.

Figure 6.14 Theoretical 'shells' that don't exist—except perhaps as antelope horns.

Other biologists think that the mutations that were needed to move into these areas of the museum were just never available. Another way of putting this view is to say that the tower block of conceivable shells that we have drawn is not, in fact, a true representation of the space of all possible shells. According to this view large areas of the tower block would not be possible even if they were desirable from a survival point of view. My own instinct favours Raup's selectionist interpretation but I don't want to pursue the matter for the moment because, in any case, I only introduced shells as an illustration of what we mean by mathematical spaces of possible animals.

I cannot leave the 'no-go' areas without briefly looking at some oddities that really do exist in the world. *Spirula* is a small, swimming cephalopod mollusc (the group that includes squids and ammonites) related to *Nautilus*. The opened-out shape of the shell betokens a high *verm* (larger than 1/*flare*), and we have already met *Spirula* in this capacity in Figure 6.4. If it is suggested that high *verm* shells like this normally don't survive because they are structurally weak, *Spirula* fits the suggestion rather well. It doesn't live inside its shell but uses it as

Figure 6.15 Real shells that are out on a limb, in unfrequented parts of the Museum of All Shells. Common spirula, *Spirula spirula*, and West Indian tube shell, *Vermicularia spirata.*

an internal flotation organ. Since the shell doesn't serve for protection, nature allowed it to follow an evolutionary trajectory into what is normally a no-go region of the Museum of All Possible Shells. It is still firmly within the cube of the museum. This may be true of the West Indian tube shell drawn in Figure 6.15. which has taken up the way of life—and the shape—of a tube worm. Go to the bottom right of Figure 6.8 and you'll at least be in the general area of the museum where the West Indian tube shell is housed. On the other hand, close relatives of this creature (and also some extinct ammonites) have a much odder, less regular shape and they certainly cannot be housed in any one part of the museum.

Not only does our three-dimensional museum ignore the fact that tube cross-sections are not necessarily circles. It also ignores the rich patterns on the surface of shells: the tiger stripes and leopard spots of Figure 6.10, the V-form calligraphy of Figure 6.4a and the whole repertoire of flutings and ridgings to be found sculpted and painted on other shells. Some of these patterns could be accommodated in our model by an instruction to the computer, as follows: as you sweep round and round, building up the expanding tube as a series of rings, make every *n*th ring thicker than the rest. Depending upon the value of *n* this rule could show itself as vertical stripes at a particular spacing on the surface of the shell. More complicated rules for the computer can generate more elaborate patterns. A German scientist called Hans Meinhardt has made a special study of such rules. The top half of Figure 6.16 shows the surface patterns on two real shells, an olive shell and a volute. Below these are the strikingly similar patterns generated by computer rules implemented by Meinhardt's computer program. You can see that his rules produce results akin to those that grow tree-like biomorphs, but he thinks of them in terms not of twigs growing

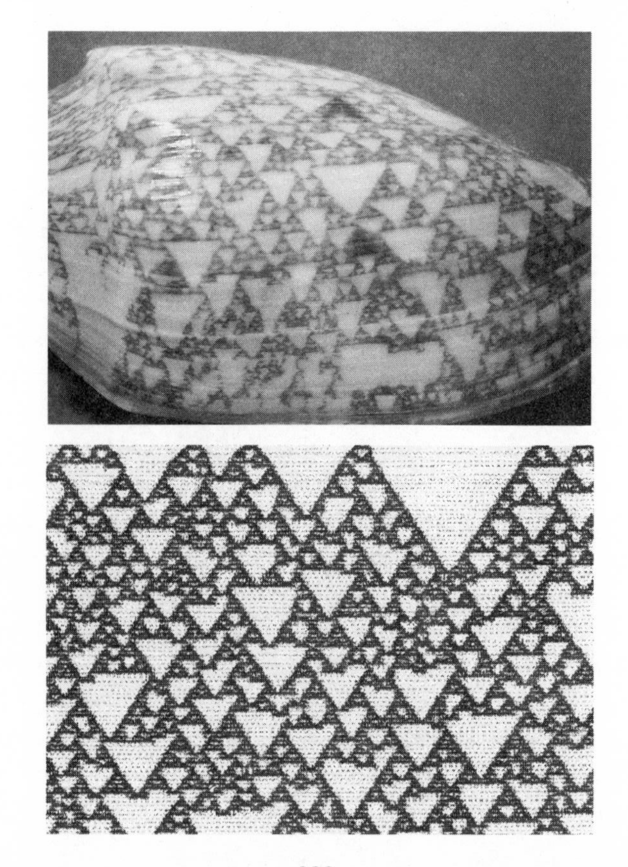

but of waves of pigment-secreting and inhibiting activity sweeping over cells. The details are found in his book, *The Algorithmic Beauty of Sea Shells*, but I must leave this subject and return to my main theme of the Museum of All Shells.

I introduced the idea of the museum because of the singular fact that—setting aside the complications of tube cross-section, ornament and variable signatures—most known variants among shells can be approximated using a mere three numbers plugged into a drawing rule. To accommodate animal forms other than shells, we shall usually

Figure 6.16 Shell surface patterns and computer-generated equivalents.

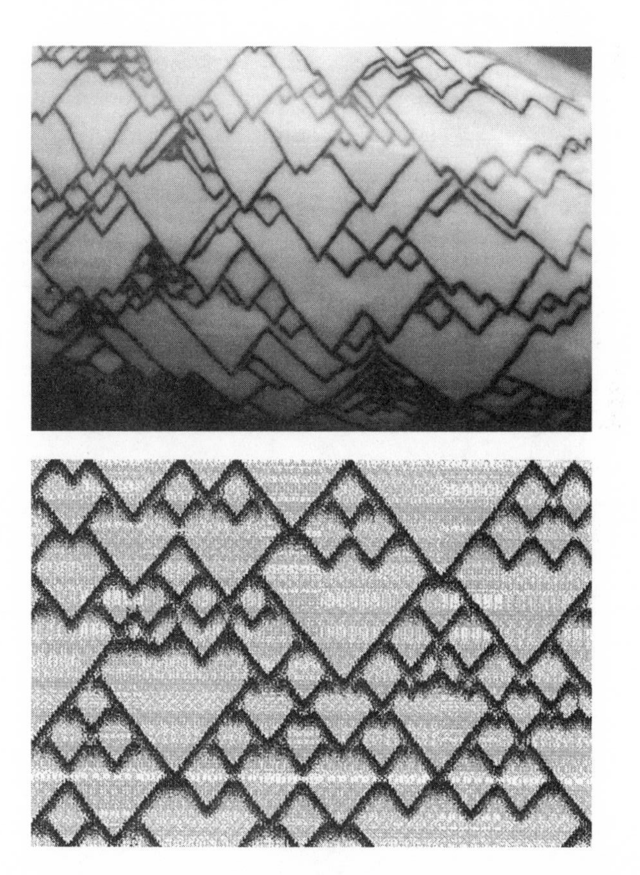

have to imagine a museum built in a larger number of dimensions than we can draw. Difficult as it is to visualize the myriad-dimensioned Museum of All Possible Animals, it is easy to keep in our heads the simple idea that animals are housed near those that they most closely resemble, and that it is possible to move in any direction, not just straight along corridors. An evolutionary history is a snaking trajectory through some part of the museum. Since evolution is going on independently in all parts of the richly diverse animal and plant kingdoms, we can think of thousands of trajectories, tunnelling in different directions through different regions of the multi-dimensional museum (notice how far we have come from the very different metaphor of Mount Improbable).

Now the controversy with which we began can be re-expressed as follows. Some biologists feel that as you walk the long corridors of the museum what you will find is smooth gradations in all directions. Large portions of the museum are never, as a matter of fact, visited by living flesh and bone but, according to this view, they would be visited if only natural selection 'wanted' to nose its way into those portions. A different set of biologists, with whom I am less in sympathy but who may be right, feel that large portions of the museum are forever barred to natural selection; that natural selection might batter eagerly on the doors of a particular corridor but never be admitted, because the necessary mutations simply cannot arise. Other parts of the museum, according to an imaginative variant of this view, far from being barred to natural selection, act like magnets or sinks, sucking animals towards them, almost regardless of natural selection's best efforts. According to this view of life, the Museum of Possible Animal Forms is not an evenly laid-out mansion of long galleries and stately corridors with smoothly changing qualities, but a set of well-separated magnets, each one bristling with iron filings. The iron filings represent animals, and the empty spaces between the magnets represent intermediate forms which might or might not survive if they came into existence, but couldn't exist in the first place. Another, and probably better, way to express this view is to say that our perception of what constitutes an 'intermediate' or a 'neighbour' in animal space is wrong. True neighbours are those forms which, as a matter of fact, can be reached in a single mutational step. These may or may not look, to our eyes, like neighbours.

I have an open mind about this controversy although I lean in one direction. On one point, though, I insist. This is that wherever in nature there is a sufficiently powerful illusion of good design for some purpose, natural selection is the only known mechanism that can account for it. I do not insist that natural selection has the keys to every corridor of the Museum of All Possible Animals, and I certainly don't think that all parts of the museum can be reached from all other parts. Natural selection is very probably not free to wander where it will. It may be that some of my colleagues are right, and natural selection's freedom of access as it snakes, or even hops, around the museum is severely limited. But if an engineer looks at an animal or organ and sees that it is well designed to perform some task, then I will stand up and assert that natural selection is responsible for the goodness of apparent design. 'Magnets' or 'attractors' in Animal Space cannot, unaided by selection, achieve good functional design. But now, let me soften my position just a little by introducing the idea of 'kaleidoscopic' embryologies.

CHAPTER 7

KALEIDOSCOPIC EMBRYOS

BODIES ARE BUILT BY THE PROCESSES OF GROWTH—mostly in the embryo—so a mutation, if it is to change the shape of a body, will normally do it by adjusting the processes of embryonic growth. A mutation might, for instance, speed up the growth of a particular piece of tissue in the embryonic head, resulting eventually in a lantern-jawed adult. Changes that occur early in foetal development can lead to dramatic knock-on effects later—two heads, perhaps, or an extra pair of wings. Such drastic mutations, for reasons that we saw in Chapter 3, are unlikely to be favoured by natural selection. In this chapter I am making a different point. This is that the *kind* of mutations that are available for natural selection to work on will depend upon the kind of embryology that the species possesses. Mammal embryology works in a very different way from insect embryology. There may be analogous, though smaller, differences in the types of embryology adopted by different orders of mammals. And the point I want to come on to is that some types of embryology may, in some sense, be 'better' at evolving than others. I don't mean more likely to mutate, which is another matter altogether. I mean that the kinds of variations thrown up by some types of embryology may be more evolutionarily promising than the kinds of variations thrown up by other types of embryology. Moreover, a form of higher-level selection—what I have previously dubbed 'the evolution of evolvability'—may lead to the world becoming peopled by types of creatures whose embryologies made them good at evolving.

Coming from a dyed-in-the-wool Darwinist like me, this might sound like the rankest heresy. Natural selection is not supposed, in nice neo-Darwinian circles, to choose among large groupings. And didn't we agree in Chapter 3 that natural selection favours a zero mutation rate (which, fortunately for the future of life, it never reaches)? How can we now claim that a particular kind of embryology might be 'good' at mutating? Well, perhaps in the following sense. Certain kinds of embryology may be prone to vary in certain ways; other kinds of embryology tend to vary in other ways. And some of these ways may be, in some sense, more evolutionarily fruitful than others, perhaps more likely to throw up a great radiation of new forms, as the mammals did after the dinosaurs went extinct. It is this that I meant when I made that rather odd suggestion about some embryologies being 'better at evolving' than others.

A fair analogy is the kaleidoscope, except that kaleidoscopes are concerned with visual beauty, not utilitarian design. The coloured chips in a kaleidoscope settle into a random heap. But, because of the cunningly angled mirrors inside the instrument, what we see through the eyepiece is an ornately symmetrical shape like a snowflake. Random taps ('mutations') on the barrel cause slight movements in the heap of chips. But we, looking through the eyepiece, see these as changes that are repeated symmetrically at all points of the snowflake. We tap the barrel over and over, and seem to wander through a minor Aladdin's cave of gaudily jewelled shapes.

The essence of the kaleidoscope is spatial repetition. Random changes are repeated at all four points of the compass. Or it may be not four but some other number of points depending on the number of mirrors. Mutations too, although they are single changes in themselves, can have their effects repeated in different parts of the body. We can regard this as another kind of non-randomness of mutation to add to the ones we dealt with in Chapter 3. The number of repetitions depends upon the type of embryology. I shall talk about various kinds of kaleidoscopic embryology. It was the experience of breeding biomorphs, and particularly the experience of putting software 'mirrors' (see below) into the Blind Watchmaker program, that led me to accept the importance of kaleidoscopic embryology. It is therefore not accidental that for purposes of illustration in this chapter I shall rely heavily on biomorphs and other computer animals.

First symmetry, and we'll begin with the lack of it. We're pretty symmetrical ourselves (though not totally so) and so are most other animals that we meet, so we're apt to forget that symmetry is not an obvious quality that every creature must have. Some groups of Protozoa (single-celled animals) are asymmetrical: cut them any way you wish, and the two bits won't be identical or mirror images of each other. What will be the impact of mutation on a purely asymmetrical animal? To explain this, it is easiest to switch to computer biomorphs.

The four biomorphs of Figure 7.1a are all mutant variants of the same form, and all are produced by an embryology that has no constraints of symmetry. Symmetrical shapes are not forbidden but there is no particular eagerness to produce them. Mutations just change the shape, and that is all there is to it: no 'kaleidoscopic' effects or 'mirrors' are in evidence. But look at some more biomorphs (Figure 7.1b). These are again mutant forms of one another, but their embryology has a built-in symmetry rule: the program has been modified to include a 'software mirror' down the midline. Mutations can alter all sorts of things, just as they could for the asymmetrical biomorphs, but any random change to the left side will be mirrored on the right side as well. These forms look more 'biological' than the asymmetrical forms of the previous picture.

You can think of a symmetry rule in embryology as a restriction or 'constraint'. So it is, in the strict sense that the unconstrained embryology is theoretically capable of producing symmetrical forms as well as asymmetrical ones and is therefore more prolific of forms. But we shall see in this chapter that the symmetry constraint can turn out to be an enrichment: the very opposite of a restriction. The trouble with the unconstrained embryology is that it needs to run through a myriad of forms before, by luck, a symmetrical one will chance to occur. And even then, the long-awaited symmetry will be constantly menaced by future generations of mutation. If, no matter what else may vary, symmetry is nearly always going to be desirable, the con-

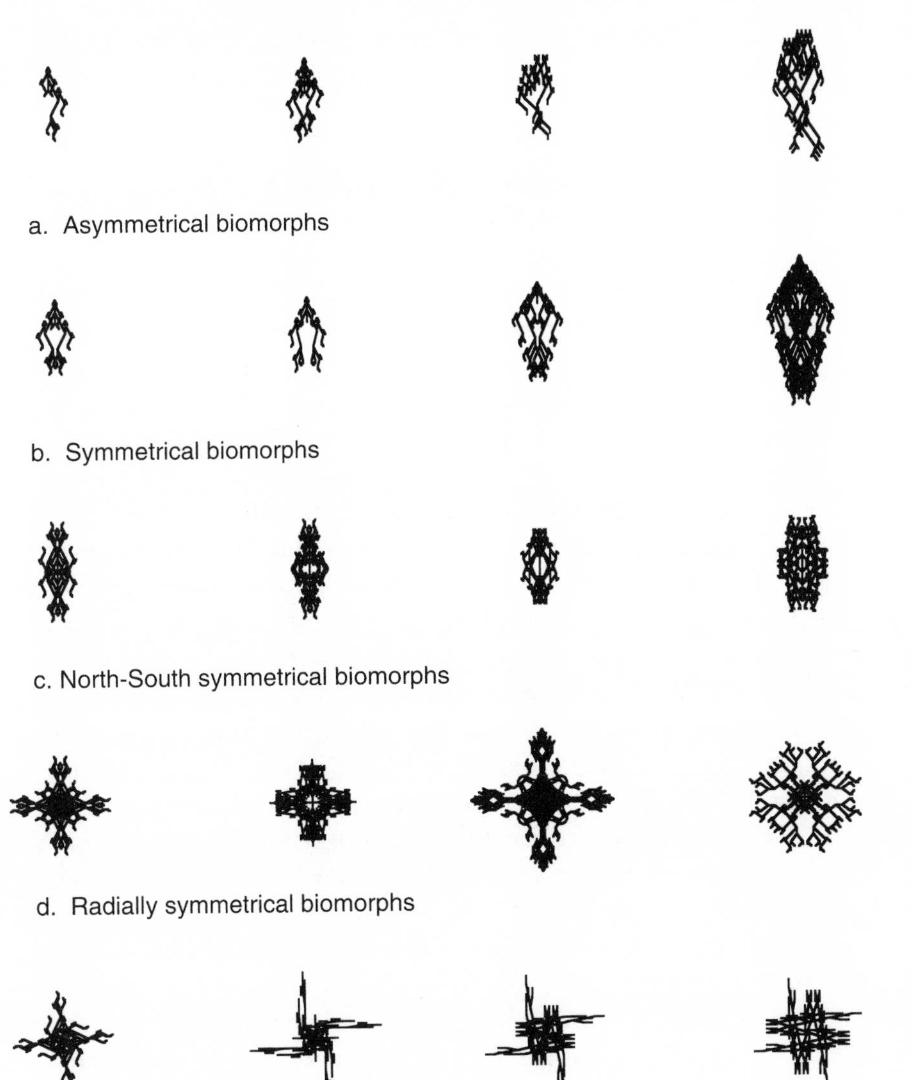

Figure 7.1 Biomorphs constrained by different numbers of 'kaleidoscopic mirrors', and therefore showing various kinds of symmetry.

strained embryology will be much more 'productive', just as it is more beautiful to our eyes. Unlike the unconstrained embryology, it won't waste time throwing up asymmetrical forms which are non-starters anyway.

As a matter of fact, the great majority of animals, ourselves included, are largely, though not entirely, symmetrical in the left–right plane. Beauty itself is not important so we must ask why, in a utilitarian sense, left–right symmetry should be a desirable characteristic. There are some zoologists wedded to an eighteenth-century idea that, with respect to major architectural features like symmetry, animals are the way they are because of some almost mystical loyalty to a 'fundamental body plan' or *Bauplan*. (*Bauplan* is just the German word for blueprint. Typically, one switches languages to indicate profundity—'these tuba notes from the depths of the Rhine', as Sir Peter Medawar sarcastically put it. But actually, if I may be allowed a parenthetic in-joke, there is irony in 'blueprint' since it suggests a 'reductionistic', one-to-one relation between plan and building which would in a genetic context offend the ideological sensibilities of the very people who are most fond of the word *Bauplan*.) I prefer the Anglo-Saxon simplicity of my colleague Dr Henry Bennet-Clark, with whom I have discussed these matters: 'All questions about life have the same answer (though it may not always be a helpful one): natural selection.' No doubt the detailed benefits of left–right symmetry vary in different animal types but he makes general suggestions too, along the following lines.

Most animals are either worm-like, or are descended from worm-like ancestors. If you think about what it is like to be a worm, it makes sense to have the mouth at one end—the end that hits food first—and the anus at the other so that you can leave waste products behind instead of inadvertently eating them. This defines a front and a rear. Then, the world usually imposes a significant difference between up and down. The minimal reason for this is gravity. In particular, many animals move over a surface such as the ground or the sea bottom. It is sensible that, for all sorts of detailed reasons, the side of the animal nearest the ground should be different from the side nearest the sky. This defines a dorsal (back) and a ventral (belly) side and therefore, given that we already have a front end and a rear end, we now also have a left and a right side. But why should the left and the right sides be mirror images of each other? The short answer is, why not? Unlike the front/rear asymmetry and the up/down asymmetry which have good justifications, there is no general reason to suppose that the best shape for a left side will be different from the best shape for a right side. Indeed, if there is a best way for a left side to be, it is reasonable to assume that the best right side will share the same qualities. More specifically, any major departure from left/right mirror symmetry may result in the animal going round in circles when it should be pursuing the shortest distance between two points.

Given that, for whatever reason, it is desirable that left sides and right sides should evolve together as lock-stepped mirror images, embryologies that are 'kaleidoscopic' with a single 'mirror' down the midline will have an advantage. New mutations that are any good will then automatically be reflected on both sides. What is the non-kaleidoscopic alternative? An evolving lineage might achieve first a beneficial change on, say, the left side of the body. Then it would have to wait through many generations of asymmetry for a matching rightside mutation to turn up. It is easy to see that a kaleidoscopic embryology might well have an advantage. Perhaps, therefore, there is a kind of natural selection in favour of kaleidoscopic embryologies of increasingly restrictive—yet correspondingly productive—character.

This is not to say that left–right asymmetries can never evolve. Mutations do occasionally arise that affect one side more strongly than the other. There are special reasons why asymmetric mutations are sometimes desirable, to fit the abdomens of hermit crabs into coiled shells, for instance, and natural selection does then favour them. We have already met flat-fish such as plaice, sole and flounders, in Chapter 4 (see Figure 4.7). Plaice have settled down on their originally left side, and the left eye has migrated round to the ancestrally right, now upper, side. Sole have done the same thing, except that they lie on their right side which may, though not necessarily, indicate that they evolved the habit independently. The plaice's ancestrally left surface has become the functionally lower, bottom-hugging skin and, appropriately, it has become flat and silvery. The ancestrally right surface has become the functionally upper, sky-pointing one and it has correspondingly become curved in shape and camouflaged in colour. The ancestral dorsal (back) and ventral (belly) sides have become the functionally left and right sides. Their respective fins, the dorsal fin and the anal fin, normally so different, have become almost exact mirror-images, as functionally left and right fins. The rediscovered left–right symmetry of plaice and sole is, in fact, a good advertisement for the power of natural selection as opposed to continentally fundamental body plans. It would be interesting (and feasible) to discover whether mutations in plaice are automatically mirrored on the (new) left and right sides (that is, the old dorsal and ventral sides). Or are they still, following the ancestral pattern, automatically mirrored on the (old) left and right sides (now lower and upper)? Has the difference between the silvery and the camouflaged side of a plaice been won in the teeth of a hostile old kaleidoscopic embryology, or with the aid of a friendly, new kaleidoscopic embryology? Whatever the answer to these questions, it serves to illustrate the point that 'hostile' and 'friendly' (to evolution) are appropriate words to use about an embryology. Once again, dare we suggest that a kind of higher-level natural selection might act to improve the friendliness of embryologies to certain kinds of evolution?

From the perspective of this chapter, the important thing about left–right symmetry is that any one mutation exerts its effects simultaneously in two places on the animal instead of one. That is what I mean by kaleidoscopic embryology: it is as though the mutations are mirrored. But left–right symmetry is not the only kind. There are other planes in which mutational mirrors might be set. The biomorphs in Figure 7.1c are symmetrical not only in the left–right plane but in the fore-and-aft plane too. It is as though there are two mirrors set at right angles. Real creatures with this 'two-mirror embryology' are harder to find than left–right symmetrical ones. Venus's girdle, a ribbon-shaped plankton swimmer of the unfamiliar phylum of ctenophores or comb jellies, is a spectrally beautiful example. More commonly, kaleidoscopic embryologies can be found which conform to four-way symmetry, like the biomorphs of Figure 7.1d.

Many jellyfish exhibit this pattern of symmetry. Members of their phylum either swim in the sea (like jellyfish themselves) or are moored to the bottom (like sea anemones), so they are not subject to the fore-and-aft pressures that we discussed for crawling animals like worms. They have every reason to possess an upper and a lower side, but they lack the pressure for a front and rear, or for a left and right. Looked at from above, therefore, there is no particular reason for any one point of the compass to be favoured over any other and they are, indeed, 'radially symmetrical'. The jellyfish in Figure 7.2 happens to be four-way radially symmetrical, but other numbers of radii are common, as we shall see. The picture, like many in this chapter, was

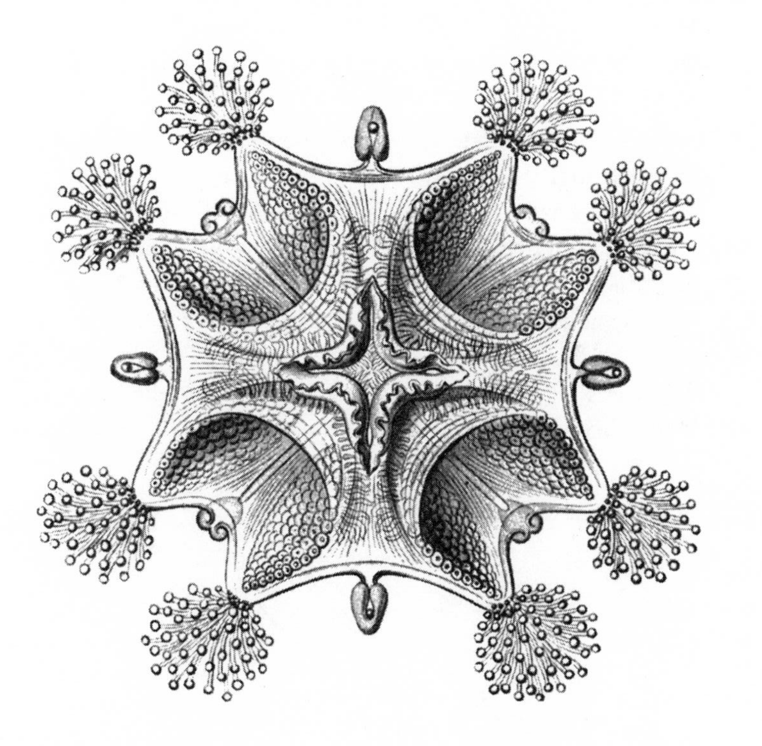

Figure 7.2 A four-way symmetrical animal: a stalked jellyfish. Note that each of the four axes is also left–right symmetrical about itself, so most variation is actually mirrored eight times.

drawn by the celebrated nineteenth-century German zoologist Ernst Haeckel, who also happened to be a brilliant illustrator.

Animals with this kind of symmetry are capable of enormous variety of form, but with a limitation which, I am again suggesting, may turn out to be not so much a limitation as a 'kaleidoscopic' enhancement. Random changes affect all four corners simultaneously. Since, at the same time, the units that are four times repeated are themselves often mirrored, each mutation is actually repeated eight times. This is very clear in the case of the stalked jellyfish in Figure 7.2 which has eight little tufts, two per corner. Presumably a mutation in tuft shape would manifest itself eight times. To see what radial symmetry would look like in the absence of this additional doubling up, look at the biomorphs in Figure 7.1e. It is quite hard to find real animals with this kind of 'swastika' or 'Isle of Man' symmetry, but Figure 7.3 shows the kind of thing we are looking for. It is the spermatozoon of a crayfish.

Most radially symmetrical animals, however many radii they may have, add left–right mirror symmetry within each radius. Therefore, from our point of view of counting the number of times a given mutation will be 'reflected', it is necessary to count the number of radii and then double it. A typical starfish, since each of its five arms is left–right symmetrical, can be said to 'reflect' each mutation ten times.

Haeckel was particularly fond of drawing single-celled organisms, such as the diatoms of Figure 7.4. Here we see kaleidoscopic symme-

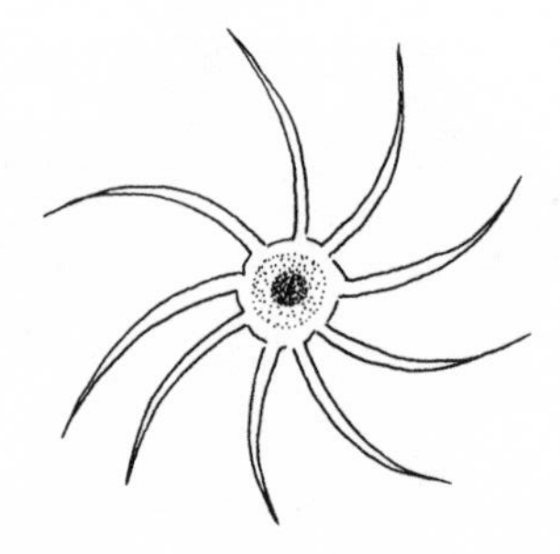

Figure 7.3 'Isle of Man symmetry': the spermatozoon of a crayfish.

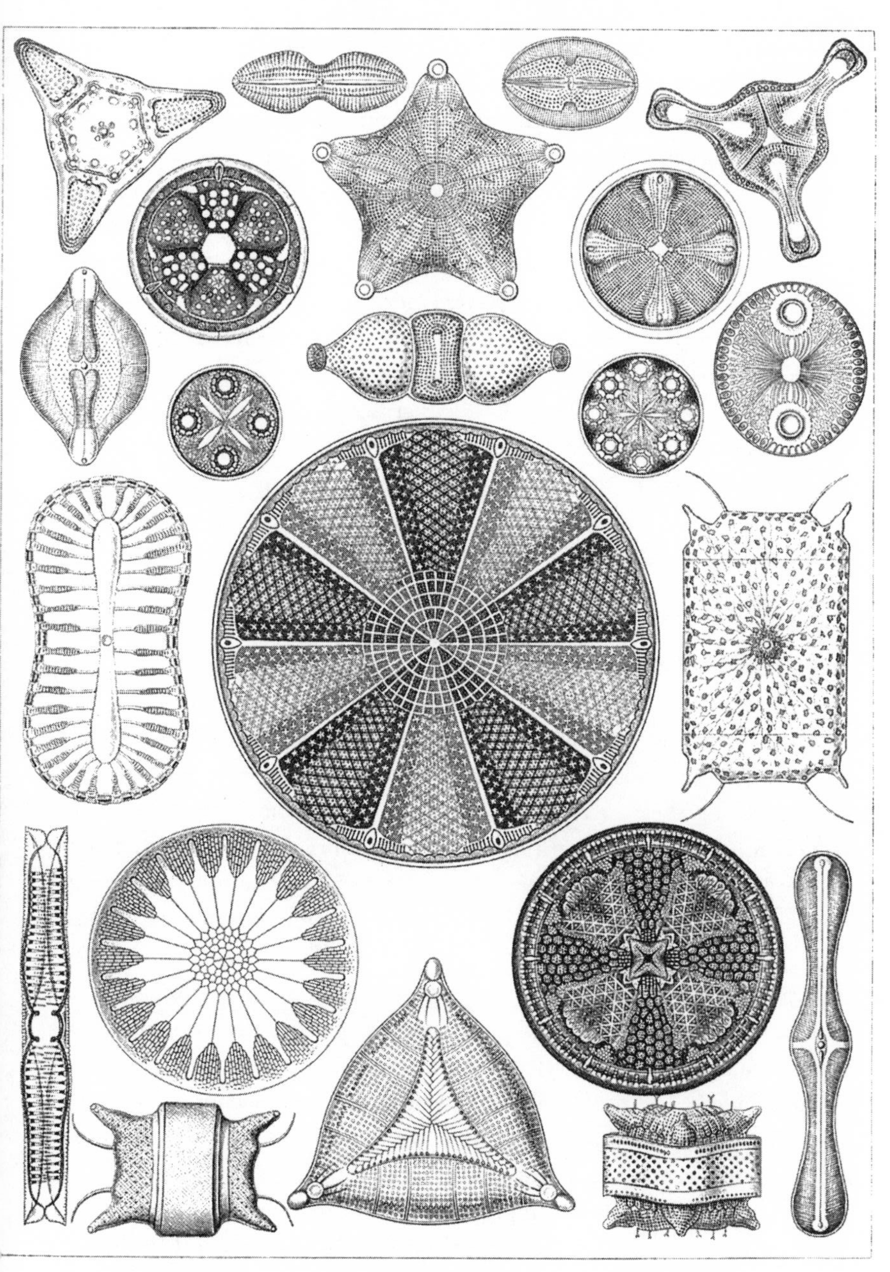

Figure 7.4 Diatoms—microscopic single-celled plants—illustrating different numbers of kaleidoscopic mirrors within one group of organisms.

tries with two, three, four, five and more 'mirrors', in addition to the left–right mirror within each arm. For each kind of symmetry, the embryology is such that mutations act not in one place but in some fixed number of places. For example, the five-pointed star near the top of Figure 7.4 might mutate to produce sharper points. In this case all five points would go sharper simultaneously. We wouldn't have to wait for five separate mutations. Presumably the different numbers of mirrors are themselves (much rarer) mutations of one another. Perhaps a three-pointed star might occasionally mutate into a five-pointed star, for example.

For me, the champions of all microscopic kaleidoscopes are the Radiolaria, another planktonic group to which Haeckel paid special attention (Figure 7.5). They too illustrate beautiful symmetries of various orders, equivalent to kaleidoscopes with two, three, four, five, six and more mirrors. They have tiny skeletons made of chalk with a beauty and elegance that has kaleidoscopic embryology written all over it.

The kaleidoscopic masterpiece in Figure 7.6 might have been designed by the visionary architect Buckminster Fuller (whom I was once privileged to hear, in his nineties, lecturing for a mesmerizing three hours without respite). Like his geodesic domes it relies for its strength on the structurally robust geometric form of the triangle. It is clearly the product of a kaleidoscopic embryology of a high order. Any given mutation will be reflected a very large number of times. The exact number cannot be determined from this picture. Other Radiolaria drawn by Haeckel have been used by chemical crystallographers as illustrations of the regular solids known since ancient times as the octahedron (eight triangular facets), the dodecahedron (twelve pentagonal facets) and the icosahedron (twenty triangular facets). Indeed D'Arcy Thompson, whom we met in connection with snail shells, would have argued that the embryologies of these exquisite Radiolarians have more in common with the growth of crystals than with embryonic development in the normal sense.

In any case, single-celled organisms such as diatoms and Radiolarians necessarily have a very different kind of embryology from many-celled ones, and any resemblance between their kaleidoscopes will

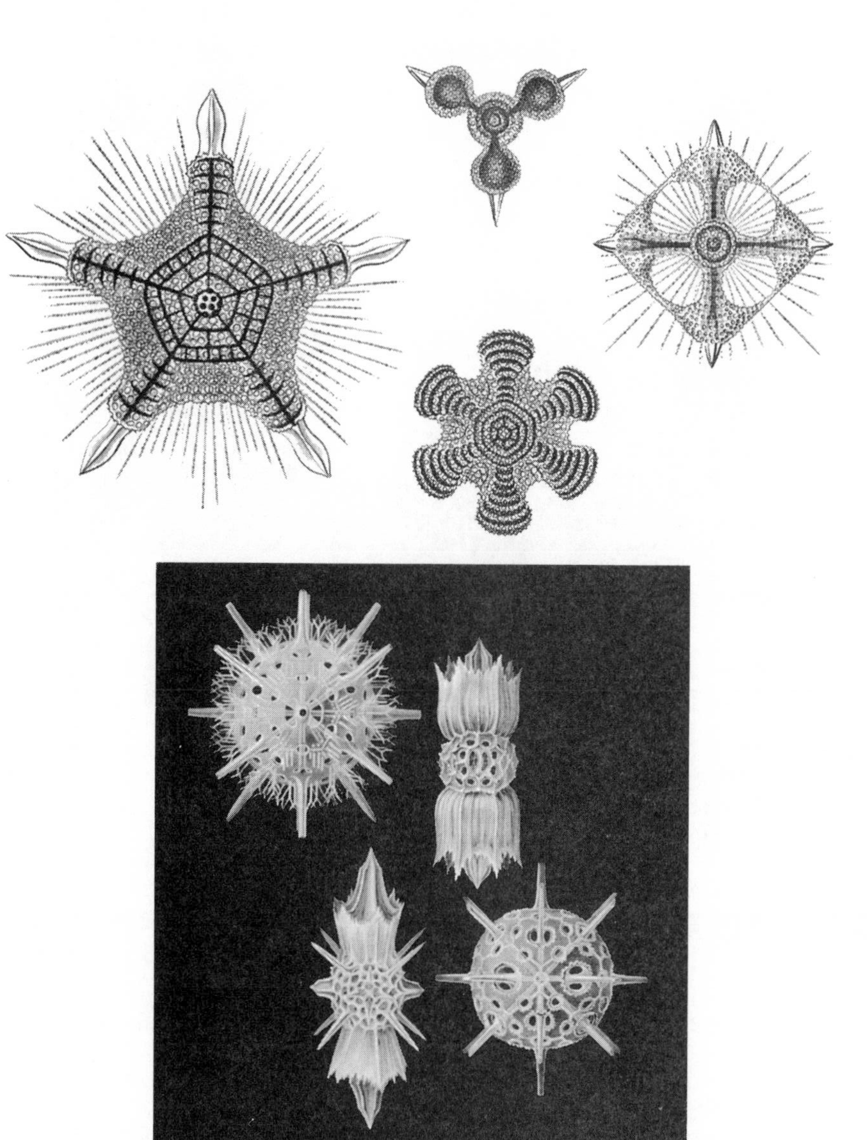

Figure 7.5 Radiolaria. More examples of different numbers of kaleidoscopic mirrors of symmetry in a group of microscopic, single-celled organisms.

Figure 7.6 A large and spectacular Radiolarian skeleton.

probably be coincidental. We've already seen an example of a four-way symmetrical many-celled animal, a jellyfish. Four, or a multiple of four, is common among such medusae, and it is presumably easy to achieve by simple duplication of some process during early embryology. There are also six-way symmetrical medusae such as those from the hydroid group known as trachymedusae (Figure 7.7).

The most famous exponents of five-way symmetry are the echinoderms—that great phylum of spiny sea creatures that includes starfish, sea-urchins, brittle-stars, sea-cucumbers and sea-lilies (Figure 7.8). It's been suggested that modern five-way symmetrical echinoderms came from three-way symmetrical remote ancestors, but they have been five-way symmetrical for more than half a billion years and it is tempting to see five-way symmetry as a central part of one of

Figure 7.7 Six-way symmetrical medusae.

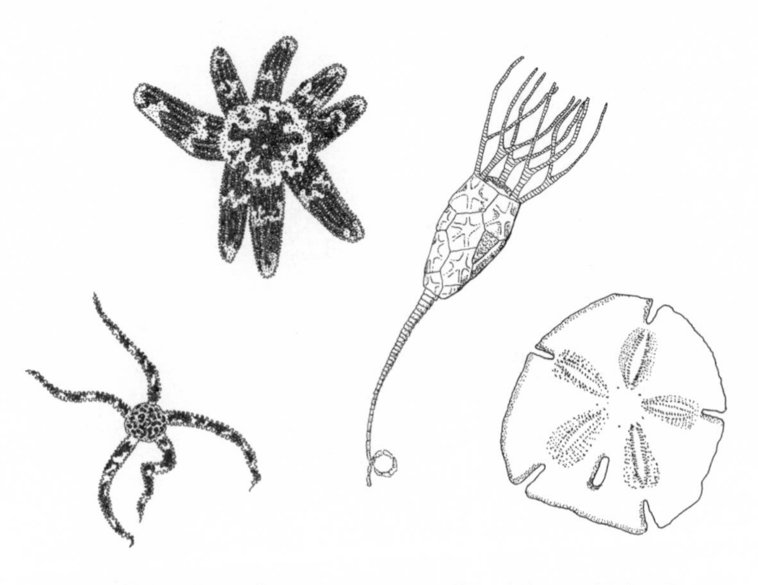

Figure 7.8 Echinoderms from various groups: (from left to right) brittle star, many-armed starfish (which has probably suffered some individual arm loss and regeneration, hence its unequal arms), sea lily, sand dollar.

those highly conserved *Baupläne* that continentally inspired zoologists are so fond of. Unfortunately for this idealistic view, not only is there a good minority of starfish species with arm counts other than five, but even within respectably five-pointed species mutant *individuals* with three, four, or six-way symmetry sometimes turn up.

On the other hand, contrary to what we might expect from our simple analysis of what it takes to be a bottom-dwelling crawler, even echinoderms that crawl are usually radially symmetrical. And they seem to take their radial symmetry seriously in the sense that they don't mind which way they walk: no one arm is privileged. At any one time a starfish will have a 'leading arm' but from time to time it changes to lead with a different arm. Some echinoderms have rediscovered left–right symmetry over evolutionary time. The burrowing heart urchins and sand dollars, whom the sand must subject to extreme pressure to streamline, have rediscovered a front–rear asymmetry and superimposed a superficial left–right asymmetry over their shape, which is recognizably based upon that of a five-radius sea-urchin.

Echinoderms are such exquisite creatures that, when I was trying to breed lifelike biomorphs with the Blind Watchmaker program, I naturally aspired to achieve their likeness. All attempts to breed five-way symmetry were doomed. The Blind Watchmaker embryology was not kaleidoscopic in the right way. It lacked the requisite number of 'mirrors'. In fact, as we have seen, some freak echinoderms depart from five-way symmetry and I 'cheated' by simulating starfish, brittlestars and urchins with even numbers of radii (Figure 7.9).

But there is no getting away from the fact—indeed it illustrates the central point of this chapter—that the present version of the Blind Watchmaker computer program is incapable of throwing up a five-way symmetrical biomorph. In order to rectify this I'd have to make a change to the program itself (a new 'mirror', not just a quantitative mutation to an existing gene) so as to allow a new class of kaleidoscopic mutations. If this were done, I feel confident that the ordinary, if rather time-consuming, processes of random mutation and selection would produce far better likenesses of most of the major groups of echinoderms. The original version of the program, as described in *The Blind Watchmaker*, was capable of producing only left–right symmetrical mutations. The present, commercially available, program's ability to produce four-way symmetrical biomorphs, and 'Isle of Man'

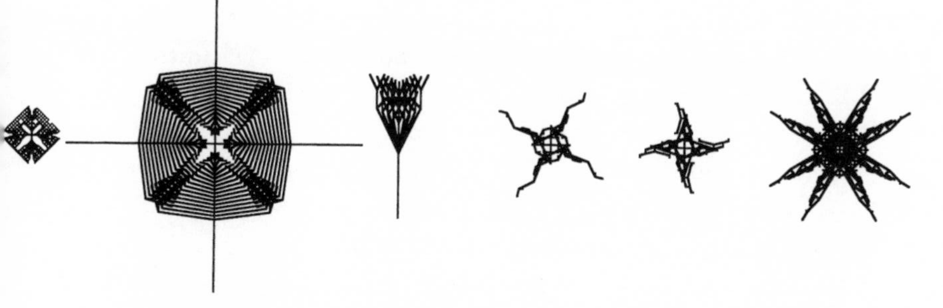

Figure 7.9 Computer biomorphs can look superficially like echinoderms, but they never achieve that elusive five-way symmetry. The program itself would have to be rewritten for that.

or 'swastika' biomorphs, results from a decision, on my part, to rewrite it so as to place a repertoire of 'software mirrors' under genetic control.

I have been talking about various kinds of symmetry as examples of kaleidoscopic embryologies. Less geometrically spectacular, but just as important in the world of real animals, is the phenomenon of segmentation. Segmentation means serial repetition as you move from front to rear of the body, usually of an ordinarily long, left–right symmetrical animal. The most obvious examples of segmented animals are annelids (earthworms, lugworms, ragworms and tubeworms) and arthropods (insects, crustaceans, millipedes, trilobites, etc.), but we vertebrates are segmented too, though in a rather different way. Just as a train is a series of trucks or carriages, each one basically like the others but differing in detail, an arthropod is a series of segments which may differ from each other in detail. A centipede is like a goods train, with all its trucks much the same as each other. You can think of other arthropods as glorified centipedes: trains with varied, special-purpose trucks and coaches (Figure 7.10).

The centipede way of organizing a body is repetitious in a simple way. There is spatial repetition all the way along the train and left–right mirroring within each segment too. But, if we move away from centipedes and their kind, there is a persistent tendency in evolution for segments to become progressively different from each other: not all mutations are simply repeated in every segment. Insects are like centipedes that have lost the legs from all segments except three: segments seven, eight and nine, counting from the front. Spiders have kept legs on four segments. Actually, both spiders and insects have kept more of their primitive 'limbs' than this. It is just that they've turned them to other uses, like antennae or jaws. Lobsters and, even more so, crabs, have carried unkaleidoscopic differentiation among segments even further.

Caterpillars have the usual three 'proper insect legs' near the front, but they have also reinvented the leg further back. These reinvented legs are squashier and otherwise rather different from the typical jointed armour legs which sprout from the three thoracic segments. Insects also typically have wings on segments seven and eight. Some

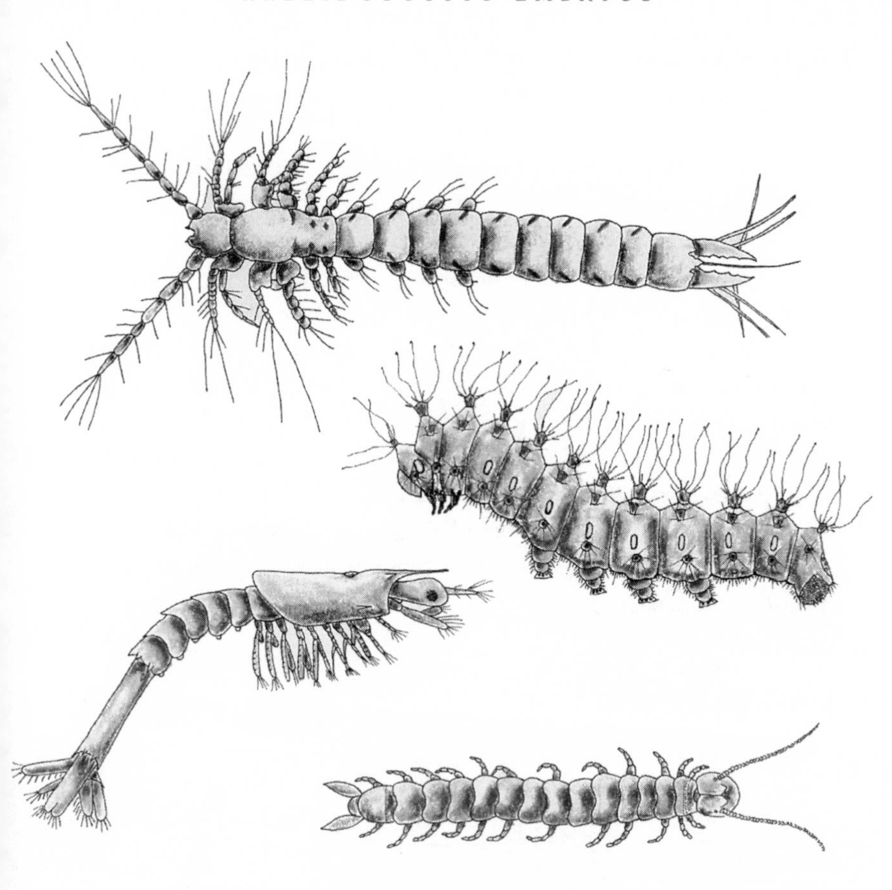

Figure 7.10 Arthropods are built up from segments repeated, often with variation, from front to back: (from top) mystacocarid crustacean, *Derocheilocaris*; giant peacock moth caterpillar, *Saturnia pyri*; dendrobranchiate shrimp, *Penaeus*; Symphyla (similar to centipedes), *Scutigerella*.

insects have no wings, and their ancestors never had them. Other insects, fleas for instance, and worker ants, have over evolutionary time lost the wings that their ancestors once had. Worker ants have the genetic equipment to grow wings: every worker could have been a queen if she had been reared differently, and queens have wings. Interestingly, a queen usually loses her wings during her own lifetime, sometimes by biting them off herself, when she has completed her mating flight and is ready to settle down underground. Wings get in the way underground, as they do where fleas live, in the thick forest that is their host's fur or feathers.

Whereas fleas have lost both pairs of wings, flies (there are lots and lots of members of the large fly family, including mosquitoes) have lost one pair of wings and kept the other pair. The second pair of wings survive in greatly reduced form as 'halteres', the tiny drumsticks sticking out just behind the working wings (Figure 7.11). You don't need to be an engineer to see that halteres would not work as wings. You need to be quite a good engineer to see what they are actually for. They seem to be tiny stabilizing instruments, doing a similar job for the insect as a gyroscope does for an aeroplane or rocket. The halteres vibrate at the wingbeat frequency. Tiny sensors at the base of the haltere detect turning forces in the three directions known to pilots as pitch, roll and yaw. It is typical of evolution to be opportunistic and make use of what is already there. An engineer designing

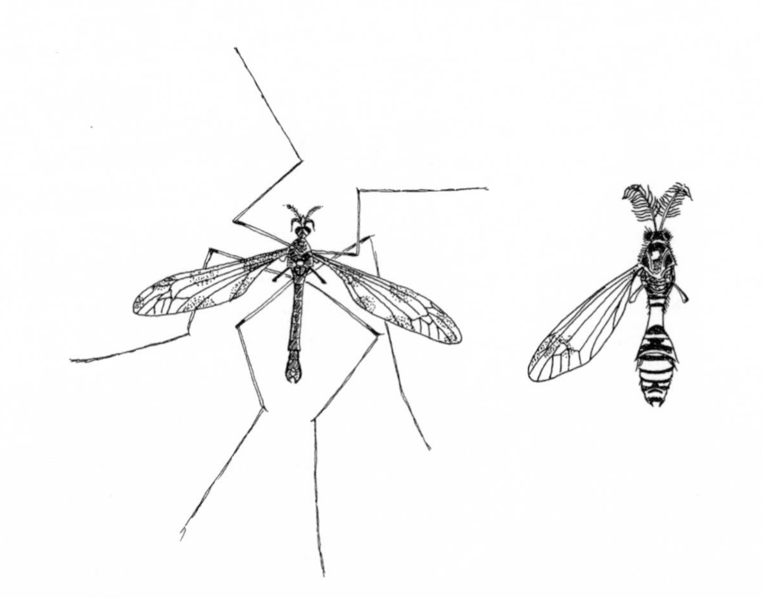

Figure 7.11 All members of the fly family have halteres in place of the second pair of wings. Large flies like these craneflies show them particularly clearly: (left) *Tipula maxima*; (right) *Ctenophora ornata* (legs and right wing not shown).

an aircraft would sit down at a drawing board and design a stabilizing instrument from scratch. Evolution achieves the same result by modifying what is already there, in this case a wing.

Just having segments evolve to be different from one another is not kaleidoscopic: quite the contrary. But there are other modes of change which we can see as kaleidoscopic, in a more sophisticated sense than we have so far met. Often arthropod bodies have a structure rather like a sentence with rounded off brackets. (If you open brackets in a sentence [inner brackets <like this> must be properly nested] the brackets eventually have to be closed again.) The phrase inside the brackets can be expanded or shortened but, no matter how long it is, every '(' must be rounded off with a ')'. The same 'proper nesting' applies to quotation marks. It also, more interestingly, applies to the embedding of subordinate clauses in sentences. 'The man who sat on a pin . . .' is like an open bracket, demanding a return to the main verb. You can say, 'The man jumped', and you can say, 'The man who sat on a pin jumped', but you cannot stop at 'The man who sat on a pin' except as the answer to a question or the caption to a picture, in which case the completion of the sentence is implicit. Grammatical sentences demand proper nesting. In a similar way, shrimps, prawns, lobsters and crayfish have six head segments all seized up together at the front end, and at the rear end a special segment called the telson. What happens in between is more variable.

We have seen one kind of kaleidoscopic mutation, the mirrored mutation reflected about various planes of symmetry. 'Grammatical' mutations would be kaleidoscopic in another sense. Once again, the variation that is permitted is constrained, but in this case not by symmetry but by rules such as: 'No matter how much variation in number of joints you allow in the middle of the leg, the leg must end with a claw.' Ted Kaehler, of the Apple Computer Company, and I collaborated to write a computer program embodying rules of this kind. It is like the Blind Watchmaker program but the 'animals' that it produces are called arthromorphs and their embryology has rules that are missing from biomorph embryology. Computer arthromorphs are a train of segmented bodies like real arthropods. Each segment has a roundish body bit—the exact shape and size is controlled by 'genes' like biomorph genes. Each segment may or may not have a jointed leg sticking out each side. That, too, is controlled by genes, and so is the thickness of the leg, the number of joints, the length of each joint and the angle of each joint. There may or may not be a claw at the end of a leg, and that too, together with its shape, is controlled by genes.

If arthromorphs had the same kind of embryology as biomorphs, there'd be a gene called *NSeg* which determined the number of segments. *NSeg* would simply have a value, which could mutate. If *NSeg* had the value eleven, the animal would have eleven segments. There'd be another gene called *NJoint* which controlled the number of joints in each limb. No matter how variable they may look—and their variety is my pride and joy—all the biomorphs in the 'safari park' of Figure 1.16 have exactly the same number of genes, sixteen. The original biomorphs of *The Blind Watchmaker* had only nine genes. The colour biomorphs have more genes (thirty-six) and the program had to be completely rewritten to accommodate them. Those are three different programs. Arthromorphs don't work like that. They don't have a fixed repertoire of genes. They have a more flexible genetic system (programming aficionados are the only readers who will wish to know that the genes of an arthromorph are stored as a Linked List with Pointers, while those of a biomorph are stored as a fixed Pascal Record). New genes can spontaneously arise in arthromorph evolution by duplication of old genes. Sometimes genes are duplicated one at a time. Sometimes they are duplicated in hierarchically structured clusters. This means that theoretically a mutant child can have twice as many genes as its parent. When a new gene, or set of genes, appears by duplication, the new genes start out with the same values as the ones from which they were duplicated. Deletion is a possible kind of mutation, as well as duplication, so the number of genes can shrink as well as grow. Duplications and deletions manifest themselves as changes in body form, and are therefore exposed to selection (artificial selection by eye, as for biomorphs). Often a change in the number of genes shows itself as a change in number of segments (Figure 7.12). It can also show itself as a change in the number of joints in a limb.

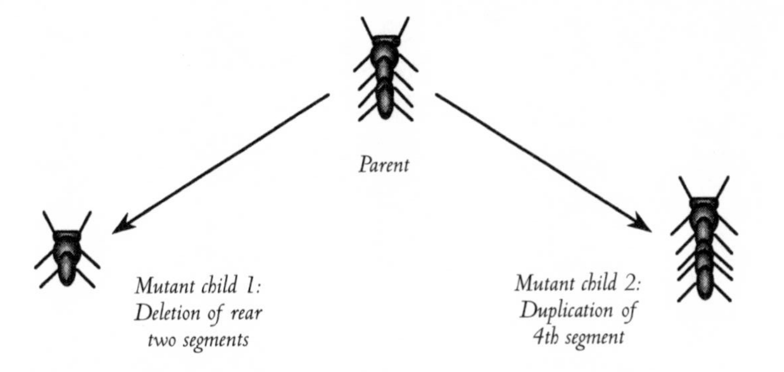

Figure 7.12 Arthromorphs differing in numbers of segments. The parent at the top has two mutant offspring.

In both cases, what emerges is a 'grammatical' tendency for middle trucks to drop in and out of the train, while leaving the front and rear trucks intact.

Duplication or deletion of segments can occur in the middle of an animal, not just at its end. And duplication or deletion of joints can occur in the middle of a limb, not just at its end. This is what gives arthromorph embryology its 'grammatical' quality: its ability to delete, or incorporate, the equivalent of a whole relative clause or prepositional clause, in the middle of a larger 'sentence'. Apart from their property of 'grammatical' nesting, arthromorphs have an additional flavour of kaleidoscopic embryology. Each quantitative detail of an arthromorph's body (for example the angle of a given claw, or the trunk width of a given segment) is influenced by three genes which multiply their numerical values together in a way that I'll explain presently. There is a gene specific to the segment concerned, a gene that applies to the whole animal, and a gene that applies to a sub-sequence of segments called a tagma. Tagma (plural tagmata) is a word that comes from real biology. Examples of tagmata in real animals are the thorax and the abdomen of insects.

For any particular detail, such as claw angle, the three genes that combine to affect it are as follows. First, the gene that is peculiar to the individual segment. This is not kaleidoscopic at all, for when it mutates it affects only the segment in question. Figure 7.13a shows an arthromorph in which every segment has a different value of the

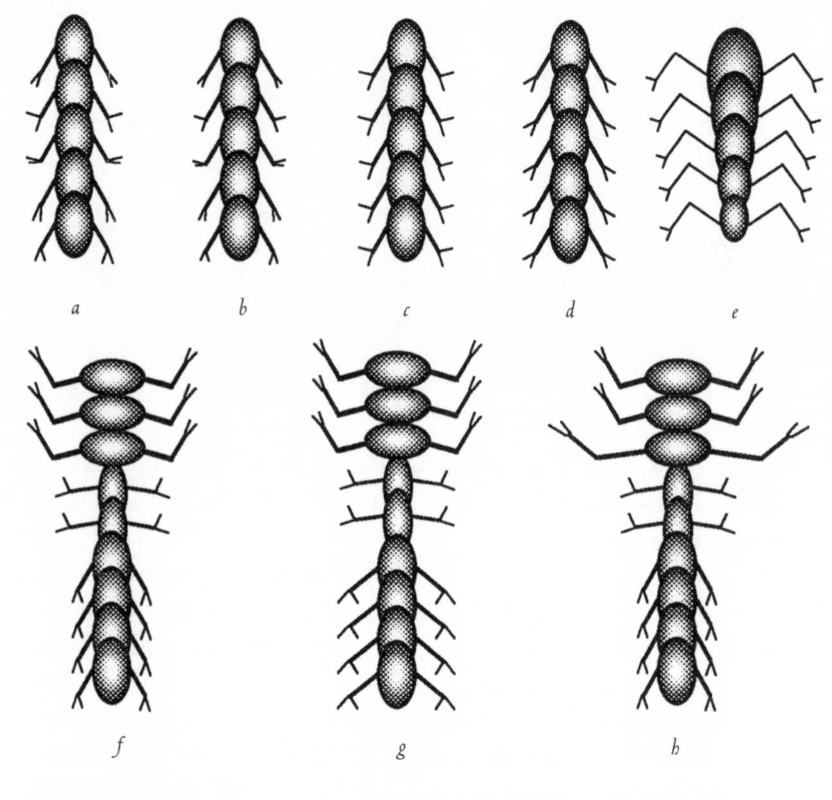

Figure 7.13 Arthromorphs chosen to illustrate various kinds of genetic effect: *(a)* arthromorph with different claw-angle gene for every segment; *(b)* mutation of whole body gene for claw angle; *(c)* arthromorph with no variation among segments; *(d)* same as *(c)* but with a single mutation affecting the whole body gene for claw angle; *(e)* gradient of segment sizes, not affecting limbs; *(f)* arthromorph with three tagmata, differing in several features but uniform within each tagma; *(g)* same as *(f)* but with a mutation affecting limbs at the level of the (third) tagma; *(h)* mutation of *(f)* affecting the limbs in one segment only.

segment-level gene for claw angle. The result is that each segment has a different claw angle. In all arthromorphs, by the way, there is simple left–right symmetry.

Moving to the second of the three genes that affects, say, the claw angle, this is the one that influences all the segments of the entire animal. When it mutates, the claws in all the segments simultaneously change, right along the length of the animal. Figure 7.13b shows an arthromorph which is the same as Figure 7.13a except that the claws are slightly pulled in—shortened. The gene affecting claw size at the level of the whole animal has mutated to a smaller value. The result is that the individual claws shrink, while retaining their segment-level peculiarities relative to one another. Mathematically, as I said, this effect is achieved by multiplying the numerical value of each individual segment-level gene for claw angle by the numerical value of the whole body-level gene for claw angle. Claw angle, of course, is only one of many quantitative details being simultaneously determined by similar multiplication sums all along the 'train'. There are whole body genes affecting, say, leg length, and these multiply their values with the segment level genes for leg length. Figure 7.13c and d show arthromorphs which have no variation among segments, but which differ from each other at the level of the whole organism gene for claw angle.

The third class of gene affects a discrete region of the body, a tagma like the thorax of an insect. Whereas insects have three tagmata, arthromorphs evolve to have any number, and each tagma can have any number of segments: changes in both segment numbers and tagma numbers are themselves subject to mutation in the 'grammatical' way we've already discussed. Each tagma has a set of genes that affect the shape of the body and the limbs and claws within that tagma. For example, each tagma has a gene that affects the angle of all the claws within that tagma. Figure 7.13f shows an arthromorph with three tagmata. Most things vary between tagmata more than within any one tagma. The effect is achieved by multiplication of gene values, in the same way as we have already seen for whole body genes.

To summarize, the final size of each attribute, say claw angle, is arrived at by multiplying the numerical values of three genes: the segment gene for claw angle, the tagma gene for claw angle, and the whole organism gene for claw angle. Since multiplication by zero yields zero, it follows that if, say, the value of a gene for limb size in a given tagma were zero, the segments of that tagma would have no limbs at all, like a wasp's abdominal segments, regardless of the gene values at the other two levels. Figure 7.13g shows a daughter of the arthromorph in Figure 7.13f, which has mutated a gene for limb size at the level of the third tagma. Figure 7.13h is another daughter of the arthromorph of Figure 7.13f, but in this case a limb gene belonging to only one segment has mutated.

Arthromorphs, then, have a kind of three-tiered kaleidoscopic embryology. They may mutate within one segment, in which case the change is mirrored only once, on the opposite side of the body. They are also kaleidoscopic at the 'centipede' or whole-organism level: a mutation at this level is spatially repeated down the segments of the body (and is also left–right mirrored). And they are kaleidoscopic at the intermediate, 'insect' or tagma level: a mutation at this level affects all the segments in one local cluster of segments, but not in the rest of the body. I conjecture that, if arthromorphs had to make their living in the real world, their three-tiered kaleidoscopic mutations might have benefits, for the same kind of reasons grounded in evolutionary economics, as we have already discussed in the case of mirrors of symmetry. If, say, the limbs of the middle tagma of the body function as walking legs, while the limbs of the hind tagma of the body function as gills, it makes sense that evolutionary improvements should be repeated serially along the segments of one tagma but not the other: improvements for walking appendages are unlikely to benefit breathing appendages. Hence there may be advantages in possessing a class of mutations which, when they first appear, are already reflected in all the segments of a tagma. On the other hand there may be more particular benefits in making detailed, specialist adjustments to the limbs in particular segments, in which case embryologies may be favoured which have an additional tendency to throw up mutations that are only left–right mirrored. Finally there may sometimes be benefits in mutations simultaneously appearing over all the segments of the body, not completely overriding the existing variation among segments and among tagmata, but weighting them, for example by multiplication.

As a biologically inspired afterthought, Ted Kaehler and I introduced 'gradient' genes into our arthromorph program. A gradient gene sees to it that a particular quality of an arthromorph, such as the claw angle, is not fixed as you move from front to rear of the animal but progressively increases (or decreases). Figure 7.13e shows an arthromorph with no variation among segments apart from a (negative) gradient of segment sizes. The body tapers from front to rear.

Arthromorphs breed, and evolve by artificial selection, in the same kind of way as biomorphs. A parent arthromorph sits in the centre of the screen, surrounded by its randomly mutated offspring. As in the case of biomorphs, the human selector sees no genes but only their consequences—body shapes—and chooses which will breed (and once again there is no sex). The chosen arthromorph glides to the centre and surrounds itself with a litter of its mutant progeny. As the generations go by, changes in numbers of genes, and changes in values of genes, take place behind the scenes by random mutation. All that the human chooser sees is a gradually evolving sequence of arthromorphs. Just as all computer biomorphs can be said to have been bred from 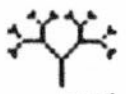 so all arthromorphs can be said to have been bred from  . The neat shading of the body segments to make them look solid is a cosmetic touch which does not vary in the existing program, though it could easily be brought under (three-tiered) genetic control in future versions of the program. Comparable to Figure 1.16's biomorph safari park, Figure 7.14 is a zoo of arthromorphs that I have bred from time to time by artificial selection, usually choosing in favour of some sort of biological realism.

This zoo includes forms that vary at all levels of the kaleidoscopic embryology. You can recognize creatures with tapering bodies as having at least one gradient gene. You can recognize clear-cut division into tagmata: groups of neighbouring segments resemble each other more than they resemble other segments. But you can still recognize some variation in form even among the segments within a tagma. Real insects, crustaceans and spiders vary in similarly tiered kaleidoscopic ways. Especially revealing are the so-called homeotic mutations of real arthropods, mutations that cause one segment to change so that it follows the pattern of development normal for a different segment.

Figure 7.15 shows examples of so-called homeotic mutations in the fruitfly *Drosophila* and in the silkworm caterpillar. The normal *Drosophila*, like all flies, has only a single pair of wings. The second

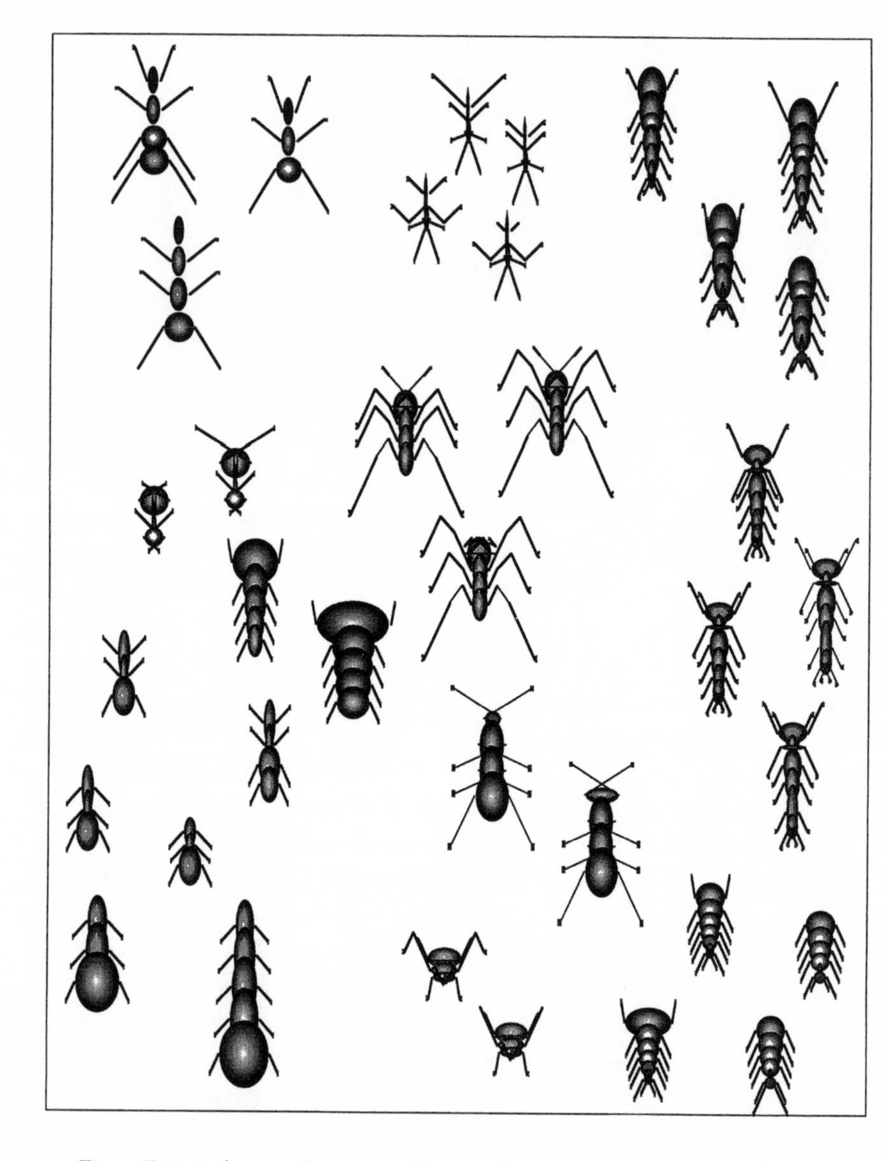

Figure 7.14 Arthromorph zoo. A collection of arthromorphs bred by artificial selection with an eye to their resemblance, however vague, to real arthropods.

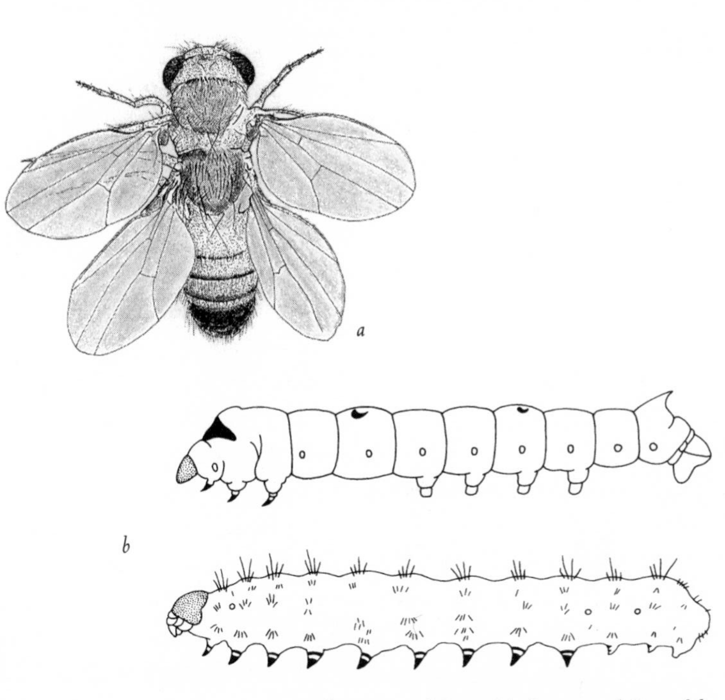

Figure 7.15 Homeotic mutations: *(a)* four-winged *Drosophila*. In normal *Drosophila* the second pair of wings is replaced by halteres, as in Figure 7.11; *(b)* normal (upper) and mutant (lower) silkworm caterpillars. Normally there are proper insect legs only on the three thoracic segments. The mutant has nine 'thoracic' segments.

pair of wings is replaced by halteres as explained above. The picture shows a mutant *Drosophila* in which not only is there a second pair of wings instead of halteres, the entire second thoracic segment is reduplicated in substitution for the third thoracic segment. In arthromorphs this effect would be achieved by a 'grammatical' duplication followed by deletion. Figure 7.15b shows a mutant silkworm caterpillar. Normal caterpillars have three 'proper' jointed legs like any other insect, although, as I said, their rear segments have squashier, 'reinvented' legs. But the mutant caterpillar at the bottom of Figure 7.15 has nine pairs of 'proper' jointed legs. What has happened is that segments of the thoracic tagma have been duplicated, just as in the right-hand arthromorph of Figure 7.12. The most famous homeotic mutation is 'antennapedia' in *Drosophila* fruitflies. Flies with this mutation have a normal-looking leg poking out of the socket where an antenna ought to be. The legproducing machinery has been switched on in the wrong segment.

Such mutants are pretty freakish and unlikely to survive in nature, which is another way of saying that evolution is unlikely to incorporate homeotic mutations. I therefore did a double-take when I was walking (rather hurriedly, for I am squeamish about food) past a table piled high with leggy seafood at a banquet in Australia. The animal that caught my attention is known to Australian gourmets as a deepwater bug. It is a kind of lobster, member of a group known variously around the world as slipper lobsters, Spanish lobsters and shovelnosed lobsters. Figure 7.16 shows a typical specimen, borrowed from the Oxford Museum, of the genus *Scyllarus*. What strikes me about these animals is that they appear to have two rear ends. The illusion results from the fact that the antennae at the front (strictly the second pair of antennae) look just like the appendages called uropods which are the most prominent feature of the rear end of any lobster. I do not know why the antennae are this shape. It may be that they are

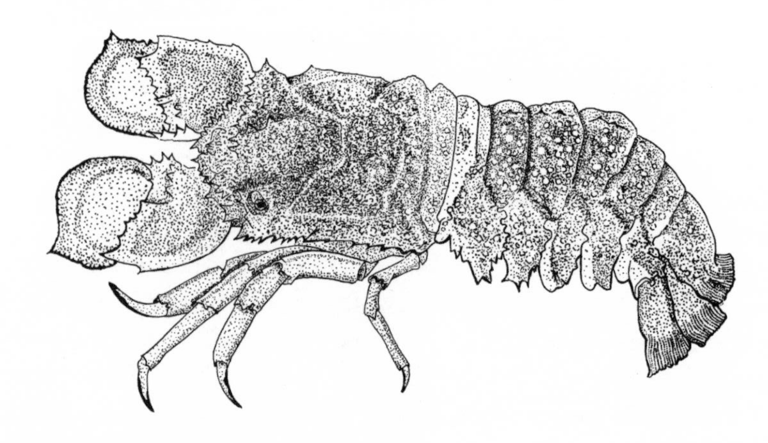

Figure 7.16 Could this represent a homeotic mutant that has been evolutionarily successful in nature? *Scyllarus*, a shovelnosed lobster.

used as shovels, or it may be that predators are fooled by the same illusion as impressed me. Lobsters have an exceedingly fast withdrawal reflex, using a giant nerve cell dedicated to the purpose. They shoot backwards with bewildering speed when threatened. A predator might anticipate this reflex by aiming behind the lobster's present position. This might pay off with an ordinary lobster, but with *Scyllarus* the apparent 'behind' could well turn out to be the front, and the second-guessing predator would have pounced in precisely the wrong direction. Whether or not this particular speculation is justified, these lobsters presumably derive some benefit from their oddly shaped antennae. Whatever that benefit is, I shall press these animals into the service of a more foolhardy speculation. My suggestion is that *Scyllarus* may actually present an example in the wild of a homeotic mutation, analogous to antennapedia in *Drosophila* in the laboratory. Unlike antennapedia, this mutation has been incorporated into an actual evolutionary change in nature. My tentative conjecture is that an ancestral Scyllarid mutated homeotically, slipping the developmental sub-routine appropriate to a uropod into a segment where an antenna ought to be, and that the change conferred some benefit. If I am right, it would constitute a rare example of a macro-mutation's being favoured by natural selection: a rare vindication of the so-called 'hopeful monster' theory that we met in Chapter 3.

That is all very speculative. Homeotic mutations definitely occur in the laboratory, and embryologists draw upon their inspiration to build a detailed picture of the mechanics of development of the segmental body plan of arthropods. Fascinating as these details are, they are beyond the scope of this chapter. I'll conclude by inviting the reader to contemplate some real arthropods in the light of the computer arthromorphs and their three-tiered kaleidoscopic mutations.

Look at the real arthropods of Figure 7.17, and imagine how they might have evolved their form through arthromorph-style kaleidoscopic genes. Do any of them, for instance, have the tapering pattern that we saw in Figure 7.13e? Now, again looking at the real arthropods, imagine a mutation that changes some small detail of the tips of limbs, or some detail of the shape of the trunk region of segments. First think about your imagined mutation applying to only a

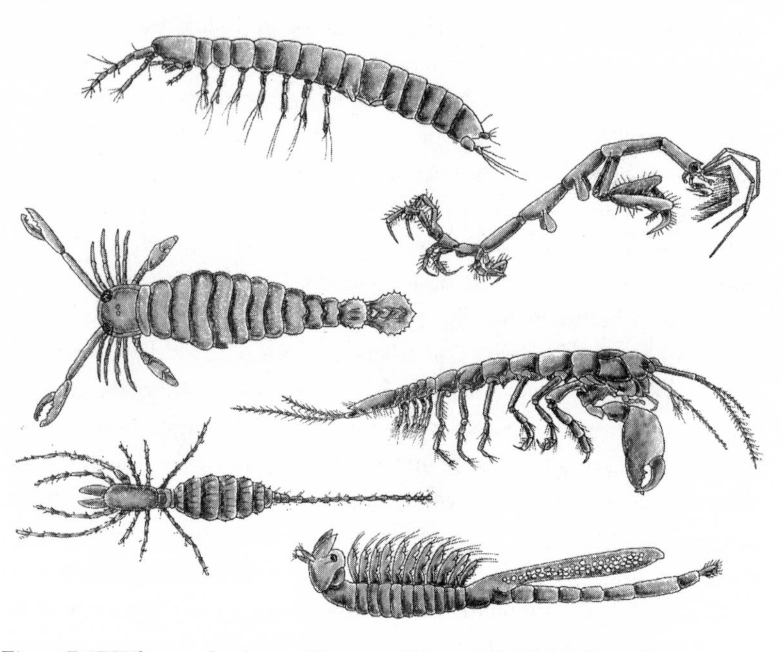

Figure 7.17 What can be done with segments: a sampling of arthropods. Clockwise from top left: four crustaceans, a palpigrade (distant relation of spiders and scorpions), and a eurypterid (extinct giant 'sea scorpion', which could easily reach a length of nearly three metres).

single segment. My guess is that you have automatically imagined it mirrored on the left and right side, but this is not inevitable. It is, in itself, an example of kaleidoscopic embryology. Now think of a mutation affecting the tips of limbs, but this time the limbs of a sequence of adjacent segments. The animals of Figure 7.17 show several examples of sequences of adjacent segments that resemble each other. Third, think of a similar mutation but affecting the tips of limbs in all segments of the body (all segments that have limbs at all, that is). I find that the experience of thinking about arthromorphs and their three-tiered kaleidoscopic embryology makes me see real arthropods, like the ones in Figure 7.17, as if through new eyes. Moreover, as in the case of mirrors of symmetry, it is easy to imagine that embryologies with arthromorph-style kaleidoscopic 'restrictions' might prove paradoxically richer in evolutionary potential than more lax, unrestricted embryologies. The shapes in Figure 7.17, and those of countless other arthropods not illustrated here, seem to me to make a special kind of sense in the light of this way of thinking.

The central message of this chapter is that kaleidoscopic embryologies, whether working through segments and clusters of segments arranged in a line from front to rear as in an insect, or through 'mirrors' of symmetry as in a jellyfish, are paradoxically both restrictions and enhancements. They restrict evolution in that they limit the range of variation available for selection to work upon. They enhance evolution in that—to put it in language that personifies selection forgivably—they save natural selection from wasting its time exploring vast regions of search space which are never going to be any good anyway. The world is populated by major groups of animals—arthropods, molluscs, echinoderms, vertebrates—each one of which has a form of kaleidoscopically restricted embryology which has proved evolutionarily fruitful. Kaleidoscopic embryologies have what it takes to inherit the earth. Whenever a major shift in kaleidoscopic mode or 'mirror' has spawned a successful evolutionary radiation, that new mirror or mode will be inherited by all the lineages in that radiation. This is not ordinary Darwinian selection but it is a kind of high-level analogy of Darwinian selection. It is not too fanciful to suggest as its consequence that there has been an evolution of improved evolvability.

CHAPTER 8

POLLEN GRAINS AND MAGIC BULLETS

I WAS DRIVING THROUGH THE ENGLISH COUNTRYSIDE with my daughter Juliet, then aged six, and she pointed out some flowers by the wayside. I asked her what she thought wildflowers were for. She gave a rather thoughtful answer. 'Two things,' she said. 'To make the world pretty, and to help the bees make honey for us.' I was touched by this and sorry I had to tell her that it wasn't true.

My little girl's answer was not too different from the one that most adults, throughout history, would have given. It has long been widely believed that brute creation is here for our benefit. The first chapter of Genesis is explicit. Man has 'dominion' over all living things, and the animals and plants are there for our delight and our use. As the historian Sir Keith Thomas documents in his *Man and the Natural World,* this attitude pervaded medieval Christendom and it persists to this day. In the nineteenth century, the Reverend William Kirby thought that the louse was an indispensable incentive to cleanliness. Savage beasts, according to the Elizabethan bishop James Pilkington, fostered human courage and provided useful training for war. Horseflies, for an eighteenth-century writer, were created so 'that men should exercise their wits and industry to guard themselves against them'. Lobsters were furnished with hard shells so that, before eating them, we could benefit from the improving exercise of cracking their claws. Another pious medieval writer thought that weeds were there to benefit us: it is good for our spirit to have to work hard pulling them up.

Animals have been thought privileged to share in our punishment for Adam's sin. Keith Thomas quotes a seventeenth-century bishop on the point: 'Whatsoever change for the worse is come upon them is not their punishment, but a part of ours.' This must, one feels, be a great consolation to them. Henry More, in 1653, believed that cattle and sheep had only been given life in the first place so as to keep their meat fresh 'till we shall have need to eat them'. The logical conclusion to this seventeenth-century train of thought is that animals are actually eager to be eaten.

> The pheasant, partridge and the lark
> Flew to thy house, as to the Ark.
> The willing ox of himself came
> Home to the slaughter, with the lamb;
> And every beast did thither bring
> Himself to be an offering.

Douglas Adams developed this conceit to a futuristically bizarre conclusion in *The Restaurant at the End of the Universe*, part of the brilliant *Hitchhiker's Guide to the Galaxy* saga. As the hero and his friends sit down in the restaurant, a large quadruped obsequiously approaches their table and in pleasant, cultivated tones offers itself as the dish of the day. It explains that its kind has been bred to want to be eaten and with the ability to say so clearly and unambiguously: 'Something off the shoulder, perhaps? . . . Braised in a white wine sauce? . . . Or the rump is very good . . . I've been exercising it and eating plenty of grain, so there's lots of good meat there.' Arthur Dent, the least galactically sophisticated of the diners, is horrified but the rest of the party order large steaks all round and the gentle creature gratefully trots off to the kitchen to shoot itself (humanely, it adds, with a reassuring wink at Arthur).

Douglas Adams's story is avowed comedy but, to the best of my belief, the following discussion of the banana, quoted verbatim from a modern tract kindly sent by one of my many creationist correspondents, is intended seriously.

Note that the banana:

1. Is shaped for human hand
2. Has non-slip surface
3. Has outward indicators of inward contents: Green—too early; Yellow—just right; Black—too late
4. Has a tab for removal of wrapper
5. Is perforated on wrapper
6. Biodegradable wrapper
7. Is shaped for mouth
8. Has a point at top for ease of entry
9. Is pleasing to taste buds
10. Is curved towards the face to make eating process easy.

The attitude that living things are placed here for our benefit still dominates our culture, even where its underpinnings have disappeared. We now need, for purposes of scientific understanding, to find a less human-centred view of the natural world. If wild animals and plants can be said to be put into the world for any purpose—and there is a respectable figure of speech by which they can—it surely is not for the benefit of humans. We must learn to see things through non-human eyes. In the case of the flowers with which we began our discussion, it is at least marginally more sensible to see them through the eyes of bees and other creatures that pollinate them.

The whole life of bees revolves around the colourful, scented, nectar-dripping world of flowers. I am not just talking about honey-bees, for there are thousands of different species of bee and they all depend utterly on flowers. Their larvae are fed on pollen, while the exclusive fuel for their adult flight-motors is nectar which is also entirely provided for them by flowers. When I say 'provided for them' I mean it in slightly more than an idle sense. Pollen, unlike nectar, is not provided *purely* for them, because the plants make pollen mainly for their own purposes. The bees are welcome to eat some of the pollen because they provide such a valuable service in carrying pollen from one flower to another. But nectar is a more extreme case. It doesn't have any other *raison d'être* than to feed bees. Nectar is manufactured, in large quantities, purely for bribing bees and other pollinators. The bees work hard for their nectar reward. To make one pound of clover honey, bees have to visit about ten million blossoms.

'Flowers,' the bees might say, 'are there to provide us bees with pollen and nectar.' Even the bees haven't got it quite right. But they are a lot more right than we humans are if we think that flowers are there for our benefit. We might even say that flowers, at least the bright and showy ones, are bright and showy because they have been 'cultivated' by bees, butterflies, hummingbirds and other pollinators. The original lecture upon which this chapter is based was called 'The Ultraviolet Garden'. This was a parable. Ultraviolet light is a kind of light that we can't see. Bees can, and they see it as a distinct colour, sometimes called bee purple. Flowers are bound to look very different through the eyes of bees (Figure 8.1). And in just the same way, the question 'What are flowers good for?' is a question that we are better off examining through the eyes of bees rather than through human eyes.

Figure 8.1 *(a)* evening primrose, *Oenothra*, photographed using visible (by humans) light; *(b)* the same, photographed by ultra-violet light (which insects can see but we can't) to show the star-shaped pattern in the centre. Presumably this pattern helps guide insects to the nectar and pollen.

'The Ultraviolet Garden' plays on the strangeness of bee vision only as a parable for changing our point of view about who or what it is that flowers—and all other living creatures—are 'for the good of'. If flowers had eyes, their view of the world might seem even odder to us than the alien ultraviolet visions of bees. How would bees appear through vegetable eyes? What are *bees* good for, from the point of view of the flowers? They are guided missiles for firing pollen from one flower to another. The background to this needs an explanation.

First, there are in general good genetic reasons for preferring cross-fertilization by pollen from a different plant. Incestuous self-mating would lose the benefits of sexual reproduction (whatever they are, which is an interesting question in itself). A tree that pollinated its female flowers with pollen from its own male flowers might almost as well not bother to pollinate at all. It would be more efficient to produce a vegetative clone of itself. Many plants of course do just this, and there is something to be said for it. But as we saw earlier there are also conditions in which there is even more to be said for reshuffling one's genes with those of another individual. It would require a massive digression to explain the detailed arguments, but there must be some substantial benefits to playing sexual roulette, otherwise natural selection wouldn't permit it to be such a driving obsession amongst almost all of animal and plant life. Whatever those benefits are, they would largely vanish if, instead of shuffling your genes with those of another individual, you simply shuffled them with a second, identical set of your own genes.

Flowers have no role in the life of their plant other than to exchange genes with another plant that has a different hand of genes. Some, like grasses, do it by wind. The air is lavishly flooded with pollen, a tiny proportion of which is lucky enough to drift on to the female parts of a flower of the same species (another proportion of it drifts into the noses and eyes of hayfever sufferers). This method of pollination is haphazard and, from some points of view, wasteful. It is often more efficient to exploit the wings and muscles of insects (or other vectors such as bats or humming-birds). This technique aims the pollen much more directly at its target, and consequently far less pollen is needed. On the other hand there has to be some expenditure

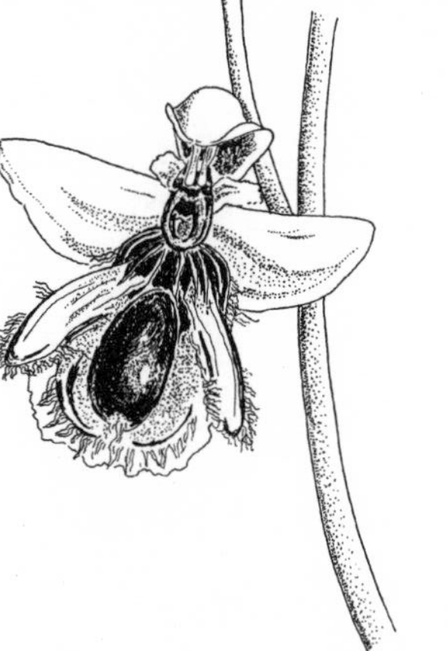

Figure 8.2 Insect-mimicking orchid. Iberian Ophrys, *Ophrys vernixia.*

on luring the insects. Part of the budget goes on advertising—bright-coloured petals and powerful scents. Part goes in bribes of nectar.

Nectar is high-quality aviation-fuel for an insect and it is costly for a plant to manufacture. Some plants duck out of the expense and employ deceptive advertising instead. Most famous are those orchids whose flowers look and smell like female insects. Male insects attempt to (Figure 8.2) copulate with the flowers and are inadvertently loaded with pollen bundles, or, at the other end of the trail, relieved of their pollen bundles. There are bee orchids that mimic female bees, and equally specialized fly orchids and wasp orchids. One of the wasp mimics, the wellnamed hammer orchid, keeps its dummy female wasp on the end of a hinged and spring-loaded stalk, cocked a fixed distance away from the pollen-bearing part of the flower (Figure 8.3). When the male wasp lands on the female dummy the spring is released. The male wasp is slammed, violently and repeatedly, against the anvil where the pollen sacs are kept. By the time the male wasp shakes himself free, his back is loaded with two pollen sacs.

Every bit as ingenious is the so-called bucket orchid, which works a little like a pitcher plant but with an important difference. The flower contains a large pool of liquid, alluringly scented to smell like the sexual attractant secreted by the females of a particular species of bee.

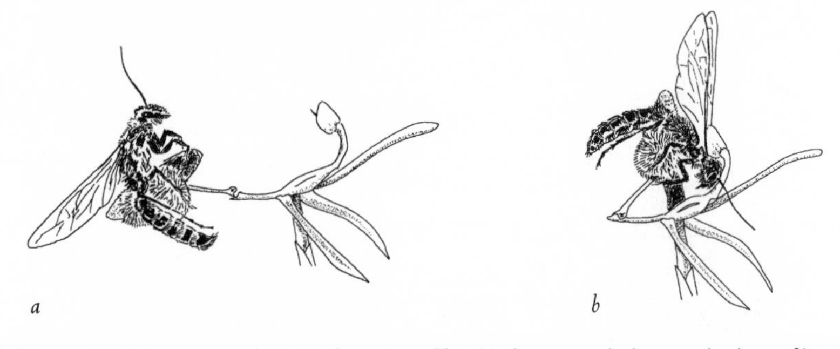

Figure 8.3 Hammer orchid, *Drakaea fitzgeraldii*: *(a)* the wasp alights on the lure; *(b)* the hinge buckles, slamming the wasp's back repeatedly against the pollinia.

A male of this species is attracted to the liquid, falls in and nearly drowns. The only escape is through a narrow tunnel. This the struggling bee eventually discovers and he crawls through it to salvation. At the far end of the tunnel there is a complicated gateway in which he is trapped for several minutes before he can wriggle free. During this final struggle at the portal of the tunnel, two large round pollen sacs are neatly transferred to his back. He then flies off and—sadder perhaps, but not wiser—falls into another bucket orchid. He again nearly drowns, again painfully pushes through the escape-tunnel and again is held for a while at the exit to freedom. During this period the second orchid relieves him of the pollen sacs and pollination is complete.

Never mind that 'sadder but not wiser'. As ever, the temptation to impute conscious intention should be resisted. It is, if anything, more tempting for the case of the plant. On both sides, the correct way to think of what is going on is in terms of unconsciously crafted machinery. Pollen that contains genes for building bee-manipulating bucket orchids is carried by bees. Pollen that contains genes for building orchids that are less accomplished at controlling bee behaviour is less likely to be carried by bees. So, as the generations go by, orchids get better at manipulating bees (although, actually, it has to be admitted that bee orchids are not in practice spectacularly successful at actually fooling bees into copulating with them).

These astonishing orchids epitomize an important aspect of pollination strategy. Many flowers seem to take great pains to get pollinated by one particular kind of animal but not any other. In the New World tropics, red tubular flowers are diagnostic of humming-bird pollination. Red is a bright and attractive colour to bird eyes (insects can't see red as a colour at all). Long, narrow tubes exclude all but specialist pollinators with long narrow beaks—humming-birds. Other flowers go out of their way to be pollinated only by bees, and we've already noted that their flowers are often coloured and patterned in the invisible (to humans) ultraviolet part of the spectrum. Yet others are pollinated only by night-flying moths. They are often white and they make use of scents in preference to visible advertisements. Perhaps the climactic stage in the progression towards an exclusive pollination partnership is the close hand-in-glove duo of fig trees with their own particular fig wasps, the example with which our book begins and ends. But why should plants be so fussy about who pollinates them?

Presumably the advantage of cultivating specialist pollinators is a more extreme version of the advantage of having animal pollinators at all, rather than wind. It narrows the target. Wind pollination is supremely extravagant, wastefully bathing the entire countryside in a rain of pollen. Pollination by jack-of-all-trades flying animals is better, but still pretty wasteful. The bee who visits your flower may fly on to a flower of a quite different species and your pollen will be wasted. Pollen borne by ordinary bees is not exactly rained over the countryside like that of a wind-pollinated grass, but it is still relatively indiscriminately splashed about. Contrast this with the bucket orchid's private species of bee, or a fig tree's private fig wasp. The insect flies unerringly, like a tiny guided missile, or like what medical journalists call a 'magic bullet', to exactly the right target from the point of view of the plant whose pollen it bears. In the case of the fig wasp, this means, as we shall see, not just another fig tree but another fig tree of precisely the right species out of the 900 fig species available. Employing specialist pollinators must permit huge savings in pollen production. On the other hand, as we shall also see, it raises other costs of its own, and it is not surprising that some plants are led by their way of life to stay with the wasteful wind as their pollinator. Other plant species are best suited by an intermediate technique along the spectrum from scattergun to magic bullet. Figs are perhaps the ultimate in dependency on the magic bullet of a particular species of pollinator and we reserve them for our climax, in the final chapter.

Returning to bees, the pollination services that they offer are truly massive. It has been calculated that, in Germany alone, honey-bees pollinate about ten trillion flowers in the course of a single summer day. It has also been calculated that 30 per cent of all human foods are derived from bee-pollinated plants, and that the economy of New Zealand would collapse if bees were wiped out. Bees, flowers might say, are put into the world to carry our pollen around for us.

The coloured and fragrant flowers of the world, then, although they may seem to be placed there for our benefit, are definitely not so. Flowers live in an insect garden, a mysterious ultraviolet garden in which, for all our vanities, we are irrelevant. Flowers have always been cultivated and domesticated but, until very recent times, the gardeners were bees and butterflies, not us. Flowers use bees, and bees use flowers. Both sides in the partnership have been shaped by the other. Both sides, in a way, have been domesticated, cultivated, by the other. The ultraviolet garden is a two-way garden. The bees cultivate the flowers for their purposes. And the flowers domesticate the bees for theirs.

Partnerships like this are quite common in evolution. There are so-called ant gardens consisting of epiphytes (plants that grow on the surface of other plants), which ants sow by bringing seeds of the right type and burying them in the soil of their nests. The plants grow out of the surface of the nest and their leaves provide food for the ants. It has been shown that some plants grow better if their roots are in an ants' nest. Other ant and termite species are specialized to cultivate fungi underground, planting the spores, weeding the gardens to rid them of competing fungus species, and fertilizing them with compost mulched from chewed-up leaves. In the case of the famous leafcutter ants of the New World tropics, all the foraging efforts of their eight-million-strong colonies are directed towards harvesting fresh-cut leaves. They can devastate an area with a ruthless efficiency reminiscent of a locust plague. Yet the leaves that they take are not to be eaten by the ants or their larvae, but are gathered purely to fertilize the fungus gardens. The ants themselves eat only the fungi, which are of a species that grows nowhere else than in the nests of this kind of ant. These fungi might say that ants are there purely to cultivate fungi, and the ants might say that the fungi exist purely to feed ants.

Perhaps the most remarkable of all the ant-loving plants are the South-East Asian epiphytes which grow a large bulbous swelling in the stem called a pseudo-bulb. The pseudo-bulb is hollowed with a labyrinth of cavities. These cavities are so like the ones ants commonly dig for themselves in soil that one would naturally suspect ants of fashioning them. This is not the case, however. The cavities are made by the plant and ants live in them (Figure 8.4).

Figure 8.4 A plant that provides custom-made accommodation for ants in return for protection. Cross-section of a pseudo-bulb of *Myrmecodia pentasperma*.

Figure 8.5 Acacia thorn. Another example of cooperation between ants and plants. These bulbous thorns are hollowed out conveniently for ants.

Better known are the species of ants that live only in special hollow thorns of acacia trees (Figure 8.5). The thorns are thick and bulbous and the plant makes them already hollowed out, apparently for no other purpose than to house ants. What the plants gain from the arrangement is protection, provided by the ants' vicious stings. This has been shown by elegantly simple experiments. Acacias whose ants have been killed by insecticide soon suffer marked increases in depredation from herbivores. Ants, if they think at all, think that acacia thorns are for the benefit of ants. Acacias think that ants are for protecting them from browsers. Should we, then, think of each member of such partnerships as working for the good of the other? It is better to think of each as using the other for its *own* good. It is a kind of mutual exploitation in which each benefits from the other enough to make the costs of helping it worth paying.

There is a temptation, for which ecologists have been known to fall, to see all of life as a sort of mutual-support encounter-group. Plants are the community's primary energy harvesters. They trap the sun's rays and make its energy available to the whole community.

They contribute to the community by being eaten. Herbivores, including the very abundant herbivorous insects, are the conduit by which the sun's energy is channelled from the primary producers, the plants, to higher stages in the food chain, the insectivores, small carnivores and large carnivores. When animals defecate or die, their vital chemicals are recycled by the scavengers such as dung beetles and burying beetles who hand the precious burden over to soil bacteria who eventually make it available to plants again.

There would not be too much wrong with this cosily benign picture of the circulation of energy and other resources, if only it were clearly understood that the participants are *not* doing it for the good of the circle. They are in the circle for the good of themselves. A dung beetle scavenges dung and buries it for food. The fact that she and her kind thereby perform a cleaning-up and recycling service which is valuable to the other inhabitants of the area is strictly incidental.

Grass provides the staple diet for a whole community of grazers, and the grazers manure the grass. It is even true that, if you removed the grazers, many of the grasses would die. But this does not mean that a grass plant exists to be eaten, or in any sense benefits by being eaten. A grass plant, if it could express its wishes, would much rather not be eaten. How, then, do we resolve the paradox that if the grazers were removed the grasses would die? The answer is that, although no plant wants to be eaten, grasses can tolerate it better than many other plants can (which is why they are used in lawns that are designed to be mown). As long as an area is heavily grazed or mown, plants that would compete with grasses cannot establish themselves. Trees cannot get a foothold because their seedlings are destroyed. Grazers, therefore, are indirectly good for grasses as a class. But this still does not mean that an individual grass plant benefits by being grazed. It may benefit from other grasses being grazed, including other plants of its own species, since this will have dividends in manure and in helping to remove competitor plants. But if the individual grass plant can get away with not being grazed itself, so much the better.

We began by lampooning the common fallacy that flowers and animals are placed in the world for the benefit of humans, cattle are docilely eager to be eaten, and so on. Marginally more defensible was the idea that they are placed in the world for the benefit of others with whom they have a naturally evolved mutualism: flowers for the benefit of bees, bees for the benefit of flowers, acacia bullhorns for the benefit of ants and their ants for the benefit of acacias. But this notion of creatures being 'for the good' of other creatures is in peril of *reductio ad absurdum.* We must have no truck with the pop ecologist's fallacy, the holisty grail of all individuals striving for the good of the community, the ecosystem, 'Gaia'. It is time to get fussy and sharpen up what we mean whenever we talk of a living creature being there 'for the benefit of' anything. What does 'for the good of' really mean? What are flowers and bees, wasps and figs, elephants and bristlecone pines—what are all living things *really* for? What kind of an entity is it whose 'benefit' will be served by a living body or a part of a living body?

The answer is DNA. It is a profound and precise answer and the argument for it is watertight, but it needs some explanation. It is this explanation that I want to come on to now and in the next chapter. I'll begin by returning to my daughter.

She was once suffering from a high fever and I suffered vicariously with her as I took my turns sitting by her bedside, sponging her down with cool water. Modern doctors could assure me that she was not in serious danger but the sleep-deprived mind of a loving father could not help recalling the countless childhood deaths of earlier centuries and the agony of each individual loss. Charles Darwin himself never recovered from the uncomprehended death of his beloved daughter Annie. The apparent injustice of her illness was said to have contributed to his loss of religious faith. If Juliet had turned to me and asked, in a piteous echo of our earlier and happier conversation, 'What are viruses for?', how should I have answered?

What are viruses for? To make us better and stronger through triumphing over adversity? (Like the 'benefits' of Auschwitz as was suggested by a professor of theology with whom I shared a debating platform on British television.) To kill enough of us to prevent the overpopulation of the world? (An especial boon in countries where effective contraception has been prohibited by theological authority.)

To punish us for our sins? (In the case of the AIDS virus, you will find plenty of enthusiasts to agree. One feels almost sorry for medieval theologians that this admirably moralistic pathogen was not around in their time.) Once again, these replies are too humancentred, albeit in a negative way. Viruses, like everything else in nature, have no interest in humans, positive or negative. Viruses are coded program instructions written in DNA language, and they are for the good of the instructions themselves. The instructions say 'Copy Me and Spread Me Around' and the ones that are obeyed are the ones that we encounter. That is all. That is the nearest you will come to an answer to the question 'What is the point of viruses?' It seems a pointless point, and that is precisely what I now wish to emphasize. I shall do so using the parallel case of computer viruses. The analogy between true viruses and computer viruses is extremely strong and it is also illuminating.

A computer virus is just a computer program, written in the same sort of language as any other computer program and travelling via the same range of media, for instance floppy discs, or the network of computers, telephone wires, modems and software that is called the Internet. Any computer program is just a set of instructions. Instructions to do what? It could be essentially anything. Some programs are sets of instructions to reckon accounts. Word processors are sets of instructions to accept typed words, move them around the screen and eventually print them. Yet other programs, like Genius 2 which recently defeated Kasparov, the Grand Master, are instructions to play chess very well. A computer virus is a program consisting of instructions that say something like this: 'Every time you come across a new computer disc, make a copy of me and put it on to that disc.' It is a 'Duplicate Me' program. It may incidentally say something more, for instance, 'Erase the entire hard disc.' Or it may cause the computer to speak, in tinny robotic tones, the words 'Don't panic'. But that is by the way. The hallmark of a computer virus, its identifying feature, is that it contains the instructions 'Duplicate me', written in a language that computers will obey.

Humans may see no reason to obey such starkly peremptory commands, but computers slavishly obey anything so long as it is written in their own particular language. 'Duplicate me' will be obeyed just as readily as 'Invert this matrix' or 'Italicize this paragraph' or 'Advance this pawn two squares'. Moreover, there is plenty of opportunity for cross-infection. Computer-users profligately exchange floppy discs, passing game programs around to friends, and useful programs too. You can easily see that, when there are lots of discs being promiscuously shared around, a program that said 'Copy me on to every disc you encounter' would spread around the world like chicken-pox. There would soon be hundreds of copies about, and the number would tend to increase. Nowadays, with information highways criss-crossing cyberspace, the opportunities for high-speed cross-infection by computer viruses are even better.

It is tempting to expostulate about the pointlessness of such parasitic programs, as I did when talking about disease viruses. What on earth is the use of a program that says nothing but 'Duplicate this program'? Admittedly it will be duplicated but isn't there something ridiculously otiose about such purely self-referential efforts? Of course there is! It is viciously futile. But it doesn't *matter* that it is futile and pointless in that sense. It can be utterly pointless and still spread. It spreads because it spreads because it spreads. The fact that it does nothing useful on the way—may even do something harmful on the way—is neither here nor there. In the world of computers and disc-swapping, it survives simply because it survives.

Biological viruses are just the same. Fundamentally a virus is just a program, written in DNA language, which is very much like a computer language even to the point of being written in a digital code. Like a computer virus, the biological virus simply says 'Copy me and spread me around'. As in the case of computer viruses we aren't suggesting that the DNA in a virus *wants* to get itself copied. It is just that, of all ways in which DNA could be arranged, only the arrangements that spell out the instructions 'Spread me' spread. The world willy-nilly becomes full of such programs. Once again, like the computer viruses, they're here because they're here because they're here. If they didn't embody instructions to ensure that they exist, they would not exist.

The only important difference between the two kinds of virus is that computer viruses are designed by the creative efforts of mischievous or evil humans, while biological viruses evolve by mutation and natural selection. If a biological virus has bad effects like sneezing or death, these are by-products or symptoms of its methods of spreading. The bad effects of computer viruses are sometimes of this type. The famous Internet Worm, which raced around the networks of the United States on 2 November 1988, had bad effects that were all non-deliberate by-products (a computer worm is technically distinct from a computer virus but the difference need not trouble us here). Copies of the program expropriated memory space and processor time, and brought around 6,000 computers to a standstill. Computer viruses, as we have seen, sometimes have bad effects which are not by-products or necessary symptoms but gratuitous manifestations of pure malice. Far from assisting the spread of the parasite these malicious effects, if anything, slow it down. Real viruses would do nothing so human-centred unless they were designed in a biological-warfare laboratory. Naturally evolved viruses don't go out of their way to kill us or make us suffer. They have no interest in whether we suffer or not. If we suffer, it is a by-product of their self-spreading activities.

'Duplicate me' instructions, like any instructions, are no use unless there is machinery set up to obey them. The world of computers is a fine and friendly place for a Duplicate Me program. Computers, linked by the Internet, abetted by people borrowing and lending discs, constitute a kind of paradise for a self-copying computer program. There is ready-made instruction-copying and instruction-obeying machinery humming and whirring and, in a sense, begging to be exploited by any program that says 'Duplicate me'. In the case of DNA viruses, the ready-made copying and obeying machinery is the machinery of cells, the whole elaborate paraphernalia of Messenger RNA, of Ribosomal RNA and of the various Transfer RNAs, each one hooking on to its own, key-coded amino acid. Never mind the details, or look them up in J. D. Watson's superbly clear *Molecular Biology of the Gene.* For our purposes it is enough to understand, first, that every cell contains a miniature analogue of a computer's instruction-obeying machinery and, second, that the machine code of all cells, in all creatures on Earth, is identical. (Computer viruses don't have that luxury, by the way: DOS viruses cannot infect Macs, and vice versa.) Computer virus instructions and DNA virus instructions are obeyed because they are written in a code that is slavishly obeyed in the environments in which they respectively find themselves.

But where does all this complaisant copying and instruction-executing machinery come from? It doesn't just happen. It has to be made. In the case of computer viruses, the machinery is made by humans. In the case of DNA viruses, the machinery is the cells of other creatures. And who manufactures those other creatures, those humans and elephants and hippos whose cells make life so easy for viruses? The answer is, other self-copying DNA manufactures them. The DNA that 'belongs' to the humans and the elephants. So, what *are* big creatures like elephants and cherry trees and mice? (I say 'big' because even a mouse, from a virus's point of view, is very very big.) And for whose benefit are mice and elephants and flowers put into the world?

We are closing in on a definitive answer to all questions of this kind. Flowers and elephants are 'for' the same thing as everything else in the living kingdoms, for spreading Duplicate Me programs written in DNA language. Flowers are for spreading copies of instructions for making more flowers. Elephants are for spreading copies of instructions for making more elephants. Birds are for spreading copies of instructions for making more birds. The cells of an elephant cannot tell whether the instructions they are slavishly obeying are virus instructions or elephant instructions. As in the case of Tennyson's Light Brigade when someone had blundered, 'Their's not to make reply, their's not to reason why, their's but to do and die'.

You will understand that I am using 'elephant' to stand for all large, autonomous creatures, for flowers or bees, for humans or cactuses, for bacteria even. The virus instructions, as we have seen, are saying 'Duplicate me'. What are the elephant instructions saying? This is the main insight that I wish to leave you with at the end of the chapter. Elephant instructions are also saying 'Duplicate me', but they are saying it in a much more roundabout way. The DNA of an elephant constitutes a gigantic program, analogous to a computer program. Like the virus DNA it is fundamentally a Duplicate Me program but it contains an almost fantastically large digression as an essential part of the efficient execution of its fundamental message. That digression is an elephant. The program says: 'Duplicate me by the roundabout route of building an elephant first.' The elephant feeds so as to grow; it grows so as to become adult; it becomes adult so as to mate and reproduce new elephants; it reproduces new elephants to propagate new copies of the original program instructions.

You can say the same about *bits* of creatures, too. The peacock's beak, by picking up food that keeps the peacock alive, is a tool for indirectly spreading instructions for making peacock beaks. The male peacock's fan is a tool for spreading instructions for making more peacocks' fans. It works by being attractive to peahens. It is good at picking up peahens while the beak is good at picking up food. Males with the most beautiful fans will have the most children to pass on copies of fan-beautifying genes. That is why peacock fans are so pretty. The fact that they are pretty to us is an incidental by-product. The peacock's fan is a gene spreader and it works via peahens' eyes.

Wings are tools for spreading genetic instructions for making wings. In the peacock's case they make their mark as gene preservers especially when the bird is surprised by a predator and shoots briefly into the air. Plants manage something akin to flight organs for their seeds (Figure 8.6), but in spite of this most people would probably not be happy to use the word 'flying', in its true sense, for plants. Plants, it seems, don't fly, and they don't have wings.

But wait! From a plant's point of view, it doesn't *need* wings of its own if it has bees' wings, or butterflies' wings, to do the job for it. In fact, I wouldn't mind *calling* the wings of a bee *plant wings*. They are organs of flight that are used, by the plant, to ferry its pollen from one flower to another. Flowers are tools for getting plant DNA into the next generation. They work like peacocks' fans, but instead of attracting peahens they attract bees. Otherwise there is no difference. Just as a peacock's fan works, indirectly, on the leg muscles of the peahen, causing her to walk towards the male and mate with him, so a flower's colours and stripes, its scent and its nectar, work on the wings of the bees and butterflies and humming-birds. The bees are drawn towards the flowers. Their wings beat and carry the pollen from one flower to

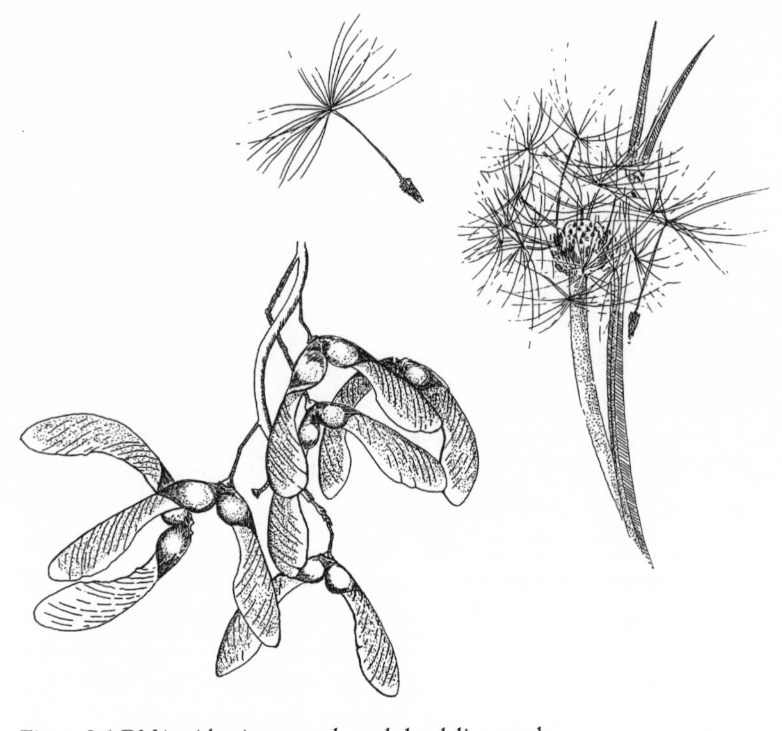

Figure 8.6 DNA with wings: maple and dandelion seeds.

another. The wings of bees can truly be called flowers' wings, for they carry flower genes just as surely as they carry bee genes.

Elephant bodies cannot tell whether they are working to spread elephant DNA or virus DNA, and bees' wings cannot tell whether they are working to spread bee DNA or flower DNA. As it happens, if we set aside exceptional cases like the bees that are fooled into wasting their time copulating with bee orchids, they are working to spread both. The difference between 'own' DNA and pollen DNA, from the point of view of the bees' executive machinery, cannot be perceived. Peacocks and bees, flowers and elephants, stand to their own DNA in much the same relation as they stand to the DNA of parasitic viruses that infest them. Virus DNA is a program that says: 'Duplicate me in a simple and direct way, using the ready-made machinery of host cells.' Elephant DNA says: 'Duplicate me in a more complicated and roundabout way that involves, first, building an elephant.' Flower DNA says: 'Duplicate me in an even more complicated and roundabout way: first, build a flower and, second, use that flower to manipulate, by indirect influences such as seductive nectar, the wings of a bee (which has already conveniently been built to the specifications of another lot of DNA, the bee's "own" DNA) to carry far and wide the pollen grains inside which are the very same DNA instructions.' We shall approach this conclusion again, from another direction, in the next chapter.

CHAPTER 9

THE ROBOT REPEATER

WE HAVE JUST CONCLUDED THAT FLOWERS AND elephants are, in effect, hosts to their 'own' DNA in the same kind of way as they are hosts to virus DNA. This is correct but it leaves difficult questions unanswered. There is an important step missing from the argument. Computer viruses have an easy time of it because the world is already full of computers, powerful, ready and waiting to obey instructions. But these computers are man-made. They are handed to the parasitic programs on a plate. DNA viruses, too, have their hosts, with all their elaborate instruction-obeying cellular machinery, handed to them on a plate. But in the case of living machinery, where does the machinery come from?

Imagine something like a computer virus which, instead of having a ready-made computer all set up to obey its instructions, had to start from scratch. It couldn't just say 'Duplicate me', because there is no computer there to obey the instructions. To be truly self-spreading, in a world without ready-provided computing and duplicating machinery, what would a self-duplicating computer program have to do? It would have to begin by saying: 'Make the machinery needed to duplicate me.' And, before that, it would have to say, 'Make the parts with which to assemble the machinery to duplicate me.' And before that, it would have to say, 'Gather the raw materials necessary to make the parts.' This more elaborate program needs a name. Let's call it the 'Total Replication of Instructions Program' or TRIP.

The TRIP has to have control over more than just an ordinary computer with a keyboard and a screen. It has to have at its disposal the equivalent of skilled hands, or grasping and manipulating devices, coupled to sensing devices in order to fashion the parts and cobble them together. The hand-like devices are necessary in order to find and assemble the parts and before that gather their raw materials. A computer can simulate things on its screen, but it cannot, on its own, build another computer like itself. To do that it needs to reach out into the real world and manipulate real, solid metals, silicon and other materials.

Let's look a little closer at the technical problems involved. Modern desktop computers can manipulate coloured shapes on cathode-ray screens, coloured pigments on printer paper and sometimes other things like sounds in stereo loudspeakers. These can all be used to create illusions of three-dimensional solidity but they really are illusions, reliant upon tricking human brains. A cube is drawn in perspective on the screen. With appropriate surface rendering it looks convincingly solid but you still can't actually pick it up and feel it, solid and weighty, between your finger and thumb. With suitable software, you could simulate cutting the cube in half and view the cross-section on your screen. But again it would not really be solid. Computers of the future may fool other senses in similar ways. Future equivalents of the computer mouse may be rigged to convey to the fingers a realistic feeling of inertia as they push a 'heavy' object around the screen. But still the object would not really be heavy, would not be made of tangible, solid stuff.

Our computer that runs the TRIP has got to manipulate more than the human imagination. It has to be capable of handling solid objects out there in the real world. How might a computer do this? It would be formidably difficult. We can begin to see this by trying to design a new kind of a computer printer, a '3-D printer'. An ordinary computer printer manipulates ink on a two-dimensional sheet of paper. One way of approximating a three-dimensional representation of, say, a cat's body, would be a set of serial sections printed on transparent sheets. The computer would laboriously slice and scan its way through the cat, from nose to tail, printing out hundreds of sheets of acetate. When the sheets are eventually stacked up into a solid block, a three-dimensional view of the cat would be visible inside the block.

This is still not a true 3-D printer, because the cat, when printed out in this way, would be embedded in a matrix of acetate. We might improve matters if we replaced the ink by a self-hardening resin. The sheets would be stacked up as before, and then dissolved or etched away leaving only the now-hardened resin. In the improbable event that the technical problems with this design could be overcome, we'd have an instrument capable of building up any three-dimensional object: a truly three-dimensional computer printer.

Our 3-D computer printer is still deeply rooted in two-dimensional preconceptions. It contrives its three-dimensional result using the principle of serial sections or slices. No output device that relied upon the serial-slice principle would be adequate for our TRIP. A useful machine, such as an internal-combustion engine, could never be made by the serial-slice technique. It needs sub-components like cylinders and pistons, flywheels and belts. These components are made of different materials from each other, and they have to be free to move relative to each other. The engine cannot be built of stacked-up slices: it must be *assembled* by bringing together previously manufactured, disjoint parts. The previously manufactured parts themselves will need assembling from smaller parts in the same way. The appropriate kind of output device for the TRIP is not a 3-D printer at all. It is an industrial robot. It has a pincer or some equivalent of a hand, capable of grasping objects. The 'hand' must be on the end of an arm-equivalent and it must have a universal joint or a set of joints capable of moving it in all three planes. It has the equivalent of sense organs, capable of guiding it towards the next object that must be picked up, and capable of steering that object towards its desired destination so that it can be fastened in position by an appropriate means.

Industrial robots of this kind do exist in modern factories (Figure 9.1). They do work, provided each one has a very particular task to perform at a particular point in an assembly line. But a normal industrial robot is still not adequate to run the TRIP program. It can put parts together—assemble them—if those parts are handed to it in a

Figure 9.1 Industrial robot from Nissan car factory. Yokohama.

fixed orientation, or regimented past it on a production line. But the whole point of our exercise is to get away from things being handed in fixed orientation, 'on a plate'. Our robot has somehow to find the raw materials for making the parts before it can begin to assemble them together. In order to do this it has to move around the world, actively seeking raw materials, mining them, gathering them up. It has to have the means to travel—something like caterpillar tracks or legs.

There are robots that do have legs, or other means of moving around the world in a quasi-purposeful way. The one in Figure 9.2 happens to be rather insect-like, except that it has four legs instead of six. It is provided with sucker feet like a fly, because its parlour trick is climbing up vertical surfaces. A favourite game of its makers is teasing it by placing a hand in just the place where the robot wants to step. The robot's foot senses the consequent unsuitability of the terrain and goes into a delightfully life-like pantomime of searching for

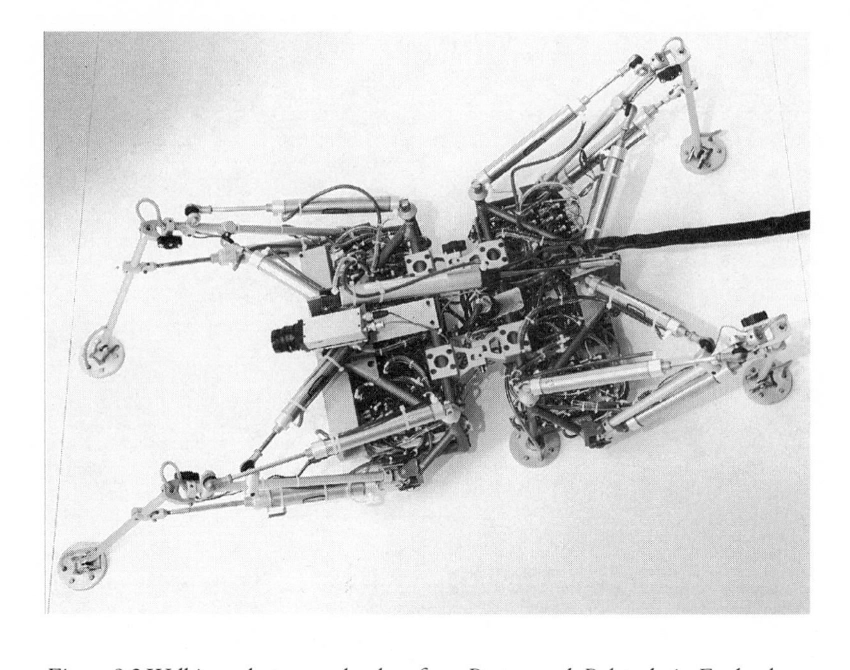

Figure 9.2 Walking robot on sucker legs from Portsmouth Polytechnic, England.

a better surface. But this is a detail of one particular robot. An earlier famous robot, the *Machina speculatrix* 'tortoise' built by W. Grey Walter of Bristol University, used to plug itself into the mains to recharge its batteries. As its batteries ran down, it manifested an increasingly restless 'appetite' for electricity and intensified its search for a mains plug. When it found one it backed on to it and stayed there until replenished. These details are not fundamental. We are talking about a machine that is capable of moving around on its own limbs and restlessly searching for something under the control of its own sense organs and its own on-board computer.

Our next task is to join the two kinds of robot together. Imagine that the walking, sucker-footed robot carries, on its back, something like the industrial, hand-wielding robot that we saw earlier. The combined machine is under the control of an on-board computer. The on-board computer has a lot of routine software for controlling the legs and the sucker feet, and for controlling the arm and hand assembly. But it is under the overall control of a master Duplicate Me program which fundamentally says: 'Walk around the world gathering up the necessary materials to make a duplicate copy of the entire robot. Make a new robot, then feed the same TRIP program into its onboard computer and turn it loose on the world to do the same thing.' The hypothetical robot that we have now worked towards can be called a TRIP robot.

A TRIP robot such as we are now imagining is a machine of great technical ingenuity and complexity. The principle was discussed by the celebrated Hungarian-American mathematician John von Neumann (one of two candidates for the honoured title of the father of the modern computer—the other was Alan Turing, the young British mathematician who, through his codebreaking genius, may have done more than any other individual on the Allied side to win the Second World War, but who was driven to suicide after the war by judicial persecution, including enforced hormone injections, for his homosexuality). But no von Neumann machine, no self-duplicating TRIP robot, has yet been built. Perhaps it never will be built. Perhaps it is beyond the bounds of practical feasibility.

But what am I talking about? What nonsense to say that a self-duplicating robot has never been built. What on earth do I think that I myself *am*? Or you? Or a bee or a flower or a kangaroo? What are all of us if not TRIP robots? We are not man-made for the purpose: we have been put together by the processes of embryonic development, under the ultimate direction of naturally selected genes. But what we actually do is exactly what the hypothetical TRIP robot is defined as doing. We roam the world looking for the raw materials needed to assemble the parts needed to maintain ourselves and eventually assemble another robot capable of the same feats. Those raw materials are molecules which we mine from the rich seam of food.

Some people find it offensive to be called a robot. This is usually because they think that a robot has to be a jerky, moronic zombie with no fine control, no intelligence and no flexibility. But these are not necessary or defining properties of a robot. They just happen to be properties of some of the robots that we have built with present-day technology. If I say that a chameleon, or a stick insect or a human is a robot that carries its own programming instructions about inside it, I am not saying anything at all about how intelligent it is. An entity can be very intelligent and still be a robot. Nor am I saying anything about how flexible it is, for a robot can be very flexible. Twentieth-century people who object to being called robots are objecting to a superficial and irrelevant association of the word (like an eighteenth-century person who objects to calling a steam carriage a vehicle of transport on the grounds that it doesn't involve a horse). A robot is any mechanism, of unspecified complexity and intelligence, which is set up in advance to work towards fulfilling a particular task. The TRIP robot's task is to distribute copies of its own program about the country, together with the machinery necessary to execute the program.

The starting point of our discussion of self-copying robots was this. We decided that a simple Duplicate Me program, like a computer virus or a real DNA virus, was all very well, but it depended upon the world being very cushy—set up with machinery capable of reading and obeying the instructions. But the world is cushy like that only because somebody, or something, else has already built that instruction-obeying machinery. We've now imagined a highly sophisticated robot which is, once again, a gigantic digression on a Duplicate Me program. Instead of just saying 'Duplicate me', the program says: 'Assemble the parts and make a new version of the entire machinery needed to copy me, and then load me into its on-board computer.'

We have arrived back at the conclusion of the previous chapter. An elephant is a huge digression within a computer program written in DNA language. An ostrich is another kind of digression, an oak tree is another. And, of course, a human is another. We are all TRIP robots, all von Neumann machines. But how did the whole process start? To answer that, we have to go back a very long time, more than 3,000 million years, probably as long as 4,000 million years. In those days the world was very different. There was no life, no biology, only physics and chemistry, and the details of the Earth's chemistry were very different. Most, though not all, of the informed speculation begins in what has been called the primeval soup, a weak broth of simple organic chemicals in the sea. Nobody knows how it happened but, somehow, without violating the laws of physics and chemistry, a molecule arose that just happened to have the property of self-copying—a replicator.

This may seem like a big stroke of luck. I want to say a few things about this 'luck'. First, it had to happen only once. In this respect, it is rather like the luck involved in colonizing an island. Most islands around the world, even quite remote ones like Ascension Island, have animals. Some of these, for example birds and bats, got there in a way that we can easily understand, without postulating a great deal of luck. But other animals, like lizards, can't fly. We scratch our heads and wonder how they got there. It may seem unsatisfactory to postulate a freak of luck, like a lizard happening to be clinging to a mangrove on the mainland which breaks off and drifts across the sea. Freakish or not, this kind of luck does happen—there are lizards on oceanic islands. We usually don't know the details, because it is not a thing that happens often enough for us to have any likelihood of seeing it. The point is that it had to happen only once. And the same goes for the origin of life on a planet.

What is more, as far as we know, it may have happened on only one planet out of a billion billion planets in the universe. Of course many people think that it actually happened on lots and lots of planets, but we only have *evidence* that it happened on one planet, after a lapse of half a billion to a billion years. So the sort of lucky event we are looking at *could* be so wildly improbable that the chances of its happening, somewhere in the universe, could be as low as one in a billion billion billion in any one year. If it *did* happen on only one planet, anywhere in the universe, that planet has to be our planet—because here we are talking about it.

My guess is that life probably isn't all that rare and the origin of life probably wasn't all that improbable. But there are arguments to the contrary. One interesting example is the 'Where are they?' argument. Imagine a South Pacific race whose island is so remote that in all the oral history of the tribe no canoe has ever found another piece of inhabited land. The tribal elders speculate as to the likelihood that there is life outside the island. The 'we are alone' faction has a powerful argument in the fact that the island has never been visited. Even if the tribe's travelling is limited to canoe-range, shouldn't there be other tribes who have progressed to more advanced boats? Why have they never come?

In the case of inhabited islands on Earth, they have all been visited by now, and today there must be few people so remote that they have not seen an aeroplane. But our island planet in the universe has never, as far as we know from properly authenticated accounts, been visited. More significantly, for the last few decades we have been equipped to detect radio communications from far away. There are about a million stars within the radius that radio waves could reach in a thousand years. A thousand years is a short time by the standards of stars and geology. If technological civilizations are common, some of them will have been pumping out radio waves for thousands of years longer than we have. Shouldn't we have heard some whisper of their existence by now? This is not an argument against life of any kind existing elsewhere in the universe. But it is an argument against intelligent, technically sophisticated life being spaced densely enough to be within easy radio range of other islands of life. If life when it starts has anything other than a low probability of giving rise to intelligent life, we might take this as evidence that life itself is rare. An alternative conclusion to this chain of reasoning is the bleak proposal that intelligent life may arise quite frequently, but typically only a short time elapses between the invention of radio and technological self-destruction.

Life may be common in the universe, but we are also at liberty to speculate that it is exceedingly rare. It therefore follows that the kind of event we are seeking, when we speculate about the origin of life, could be a very very improbable event: not the kind of event that we can expect to duplicate in the laboratory and not the kind of event that a chemist will deem 'plausible'. This is an interesting paradox, spelled out in full in a chapter of *The Blind Watchmaker* called 'Origins and Miracles'. We could be actively seeking a theory with the specific property that, when we find it, we shall judge it highly implausible! Looking at the matter in one way, we might even be positively worried if a chemist manages to support a theory of the origin of life which, using ordinary standards of probability, we judge to be plausible. On the other hand life seems to have arisen during the first half billion of the Earth's 4.5 billion years; we've been here for eight parts in nine of the Earth's age and my intuition is still that the arising of life on a planet is not all that unexpected an event.

An origin of life, anywhere, consists of the chance arising of a self-replicating entity. Nowadays, the replicator that matters on Earth is the DNA molecule, but the original replicator probably was not DNA. We don't know what it was. Unlike DNA, the original replicating molecules cannot have relied upon complicated machinery to duplicate them. Although, in some sense, they must have been equivalent to 'Duplicate me' instructions, the 'language' in which the instructions were written was not a highly formalized language such that only a complicated machine could obey them. The original replicator cannot have needed elaborate decoding, as DNA instructions and computer viruses do today. Self-duplication was an inherent property of the entity's structure just as, say, hardness is an inherent property of a diamond, something that does not have to be 'decoded' and 'obeyed'. We can be sure that the original replicators, unlike their later successors the DNA molecules, did not have complicated decoding and instruction-obeying machinery, because complicated machinery is the kind of thing that arises in the world only after many generations of evolution. And evolution does not get started until there are replicators. In the teeth of the so-called 'Catch-22 of the origin of life' (see below), the original self-duplicating entities must have been simple enough to arise by the spontaneous accidents of chemistry.

Once the first spontaneous replicators existed, evolution could proceed apace. It is in the nature of a replicator that it generates a population of copies of itself, and that means a population of entities that also undergo duplication. Hence the population will tend to grow exponentially until checked by competition for resources or raw materials. I'll develop the idea of exponential growth in a moment. Briefly, the population doubles at regular intervals, rather than just adding a constant number at regular intervals. This means that there will soon be a very large population of replicators and hence competition between them. It is in the nature of any copying process that it is never quite perfect: there are random errors in duplication. Therefore there would arise varieties of the replicator in the population. Some of these variants would have lost the property of self-duplication and their particular form would not have been retained in the population. Other variants happened to have some property that caused them to be duplicated more rapidly or more efficiently. They became consequently more numerous in the population. Since they would have been competing for the same raw materials as rival replicators, as time went by the average, typical replicator type in the population would continually have been supplanted by a new and a better average type. Better at what? Better at replicating, of course. Later, this improvement would take the form of influencing other chemical reactions so as to facilitate self-replication. Eventually, the influence would have become sufficiently complicated that an observer, had there been one (there wasn't, of course, for it takes billions of years to evolve anything that you could call an observer), might have described the process as the decoding and obeying of instructions. And if that same observer were asked what the instructions mean, he would have to reply that they mean 'Duplicate me'.

There are undoubted difficulties in this story. Among them I have already alluded to the so-called Catch-22 of the origin of life. The larger the number of components in a replicator, the more likely it is that one of them will be miscopied, leading to complete malfunctioning of the ensemble. This suggests that the first, primordial replicators must have had very few components. But molecules with fewer than a certain minimum number of components are likely to be too simple to be capable of engineering their own duplication. Ingenuity has been expended on reconciling these two apparently incompatible requirements—with some success, but the argument becomes more mathematical than is suitable for this book.

The original replication machines—the first robot repeaters—must have been a lot simpler than bacteria, but bacteria are the simplest examples of TRIP robots that we know today (Figure 9.3a). Bacteria make their livings in a great variety of ways, from a chemical point of view a far wider range of ways than the rest of the living kingdoms put together. There are bacteria that are more closely related to us than they are to other, strange kinds of bacteria. There are bacteria that obtain their sustenance from sulphur in hot springs, for whom oxygen is a deadly poison, bacteria that ferment sugar to alcohol in the absence of oxygen, bacteria that live on carbon dioxide and hydrogen, giving out methane, bacteria that photosynthesize (use sunlight to synthesize food) like plants, bacteria that photosynthesize in ways that are very different from plants. Different groups of bacteria encompass a range of radically different biochemistries compared with which all the rest of us—animals, plants, fungi and some bacteria—are monotonously uniform.

Bacteria of several different kinds got together, more than a thousand million years ago, to form the 'eucaryotic cell' (Figure 9.3b). This is our kind of cell, with a nucleus and other complicated internal parts, many of them put together from intricately folded internal membranes, like the mitochondria which I briefly pointed to in Figure 5.2. The eucaryotic cell is now seen as derived from a colony of bacteria. Eucaryotic cells themselves later got together into colonies. Volvox are modern creatures (Figure 9.3c). But it is possible that they represent the *kind* of thing that went on more than a thousand million years ago, when our kind of cells first started to band together into colonies. This ganging up of eucaryotic cells was comparable to the earlier ganging up of bacteria into eucaryotic cells and the even earlier ganging up of genes into bacteria. Larger and more densely packed gangs of eucaryotic cells are called metazoan bodies. Figure 9.3d shows a comparatively small one, a tardigrade. Metazoan bodies themselves sometimes gang up into colonies that themselves behave somewhat like individuals (Figure 9.3e).

I said that an elephant was a huge digression on a Duplicate Me program, but I could have said mouse instead of elephant and huge would still have been the right word. A volvox has a few hundred cells. A mouse is a large edifice of perhaps a billion cells. An elephant is a colony of about 1,000 trillion (10^{15}) cells, and each one of those cells is itself a colony of bacteria. If an elephant is a robot carrying its blueprint about, it is an almost unthinkably large robot. It is a colony of cells but, since those cells carry copies of the same DNA instructions, they all cooperate, working together towards the same end of duplicating their separately identical DNA data.

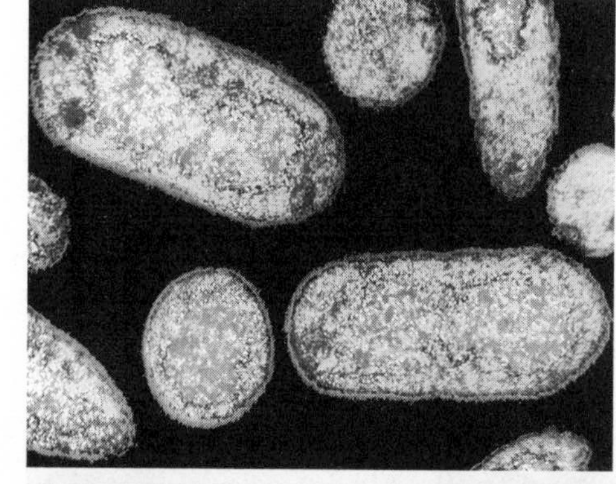

a

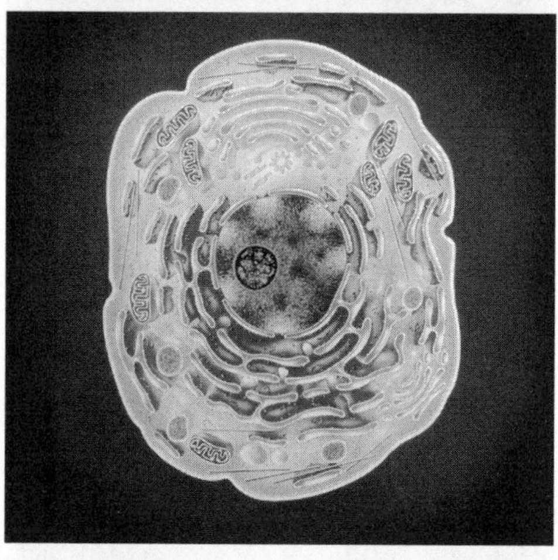

b

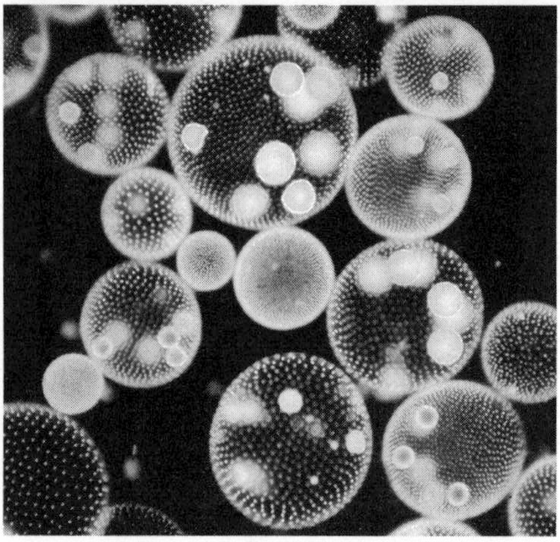

c

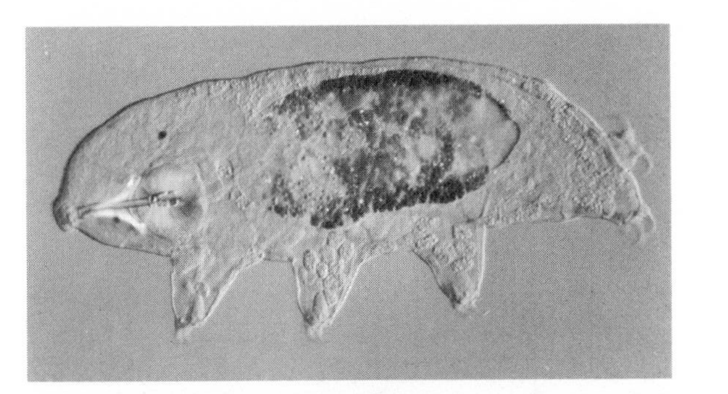

d

e

Figure 9.3 Increasing levels of organization among life forms: (a) individual bateria; (b) advanced—eucaryotic—cell with nucleus, originally evolved from a colony of bacteria; (c) volvox, a colony of differentiated eucaryotic cells; (d) a more densly packed and populous colony of differentiated eucaryotic cells, a tardigrade. A human body is another such colony—a colony of colonies, since each of our cells is a colony of bacteria; (e) a colony of individual organisms: a swarm of honey bees—a colony of colonies of colonies.

Of course an elephant isn't a particularly large thing, on any absolute scale. Compared with a star it is small. I meant large by comparison with the DNA molecules that the elephant is designed to preserve and propagate. It is large compared with the replicating elephant-makers that ride around inside it.

To get an idea of scale, imagine that human engineers built a giant mechanical robot in which they could ride, like the Greeks in their Trojan Horse. But our mechanical horse will be scaled up so that each human engineer is equivalent to one of the robot's DNA molecules in size. Remember that we are thinking of a real horse as a robot built by the genes that ride inside it. The point of the picture is that, if we built a robot horse for ourselves to ride inside, and if our robot horse were as big, relative to us, as a real horse is big, relative to the genes that built *it*, then our robot horse could bestride the Himalayas (Figure 9.4). A real, living horse is made of trillions of cells. With unimportant exceptions, every one of those cells has a full crew of genes

Figure 9.4 A horse is a robot vehicle for DNA molecules and it is very large in comparison to them. If humans built a Trojan Horse to ride inside, to a similar scale relative to ourselves, it would dwarf the Himalayas. This fantasy was painted by my mother, Jean Dawkins, for one of my Royal Institution Christmas Lectures.

riding inside it although most of them, in any one kind of cell, are sleeping.

A real living body manages to be so big (compared with the genes that built it) because it grows by a very different process from the way a man-made machine grows; quite different from the way this mechanical horse would be built, if it ever were built. The special way of growing that real living things employ is exponential growth. Another way of saying it is that living things grow by local doubling.

We start with a single cell which is very small. Or rather, it is just about the right size for the genes that make it. It is within the range that they can cope with by biochemical manipulation. Their tendrils of influence can reach all corners of a single cell, and they can fashion that single cell to have certain properties. Perhaps the most remarkable property the cell has is the ability to divide into two daughter cells more or less like itself. Being like the parent cell, each daughter cell itself is capable of dividing into two, making four granddaughter cells. Each one of the four, in turn, can double, making eight, and so on. This is exponential growth, or local doubling.

People who are not used to it find the power of exponential growth surprising. As promised, I'll spend a little time on it, because it is important. There are many vivid ways of illustrating it. If you fold a piece of paper once, you have two thicknesses. Fold it again and it is four times as thick. Another fold and you have a wad, eight layers thick. Three more folds is about as many as you can get away with, before the wad becomes too stiff to fold further: sixty-four layers thick. But suppose that this mechanical stiffness were not a problem and that you could go on folding, say fifty times. How thick would the wad of paper be then? The answer is that it would be so thick that it would reach right outside the Earth's atmosphere and beyond the orbit of the planet Mars.

In the same way, by local doubling of cells all over the developing body, the number of cells very rapidly gets up into the astronomical range. A blue whale is made of about a hundred thousand trillion (10^{17}) cells. But, such is the power of exponential growth, it would only take about fifty-seven cell generations, under ideal conditions, to produce such a leviathan. By a cell generation, I mean a doubling. Remember that numbers of cells go up 1, 2, 4, 8, 16, 32, etc. So it takes six cell generations to reach thirty-two cells. And, if you go on multiplying by two like that, it takes only fifty-seven cell generations to reach a hundred thousand trillion, the number of cells in a blue whale.

This way of calculating the number of cell generations is actually unrealistic because it gives only a minimum figure. It assumes that, after every cell generation, all the cells go on to duplicate. In fact, many cell lineages drop out of the doubling game earlier on, when they have finished building a particular part of the body, say the liver. Other cell lineages go on doubling for longer. So a blue whale in fact consists of a number of cell lineages of different length, building different parts of the whale. Some of these lineages go on dividing for more than fifty-seven cell generations. Others stop dividing after fewer than fifty-seven cell generations. In practice there are 'stem-cells', subsets of cells that are set aside for the purpose of running off copies of cells like themselves.

You can roughly calculate the minimum number of cell generations it would take, under ideal doubling conditions, to grow any animal, given its weight. You can assume that big animals don't have especially big cells, they just have more of the same kinds of cells as small animals. A naïve calculation suggests that it would take a minimum of forty-seven cell-doubling generations to grow an adult human and only about ten more cell generations to grow a blue whale. These figures are certainly underestimates, for the reason I have given. Nevertheless it remains true that, such is the power of exponential growth, you need only make a small change in how long a particular lineage of cells goes on dividing, to get a dramatic change in the final size of the bunch of cells produced. Mutations sometimes do this.

Building these colossal bodies—colossal by the standards of their DNA builders and passengers—can be called *gigatechnology*. Gigatechnology means the art of building things at least a billion times bigger than you are. The art of gigatechnology is something that our own engineers have no experience of. The biggest vehicles that we build to travel about in—large ships—are not very large relative to their builders and we can walk right round them in a matter of minutes. When we build something like a ship, we do not have the advantage of exponential building. For us, there is nothing for it but to swarm all over the structure, riveting hundreds of prefabricated steel plates together.

DNA, building its robot vehicles to ride around in, has the tool of exponential growth at its command. Exponential growth puts great power in the hands of naturally selected genes. It means that a tiny adjustment to a detail of embryonic growth control can have the most dramatic effect on the outcome. A mutation that tells a particular sub- lineage of cells to go on dividing just one more time—say go on for twenty-five cell generations instead of twenty-four—can in principle have the effect of doubling the size of a particular bit of the body. The same trick, of changing numbers of cell generations, or rates of cell division, can be used by genes during embryology to change the *shape* of a bit of the body. Modern humans have a prominent chin compared with our fairly recent ancestor *Homo erectus*. All that it takes to change the shape of the chin is small adjustments in the numbers of cell generations in particular regions of the embryonic skull.

In a way, the remarkable thing is that cell lineages stop dividing when they are supposed to, in such a way that all our bits are well proportioned relative to one another. In some cases, of course, cell lineages notoriously do not stop dividing when they should. When that happens we call it cancer. Randolph Nesse and George Williams (in their brilliant book which they wrote under the excellent title *Darwinian Medicine*, but which the publishers then saddled with an assortment of unmemorable and geographically variable titles) make a wise point about cancer. Before we wonder why we get cancer, we should wonder why we don't get it all the time.

Who knows whether humans will ever attempt to build things by gigatechnology? But people are already talking about *nanotechnology*. Just as 'giga' means a billion, 'nano' means a billionth. Nanotechnology means engineering things that are a billionth of the size of the builder.

There are people—and the leading ones are not all New Agers or

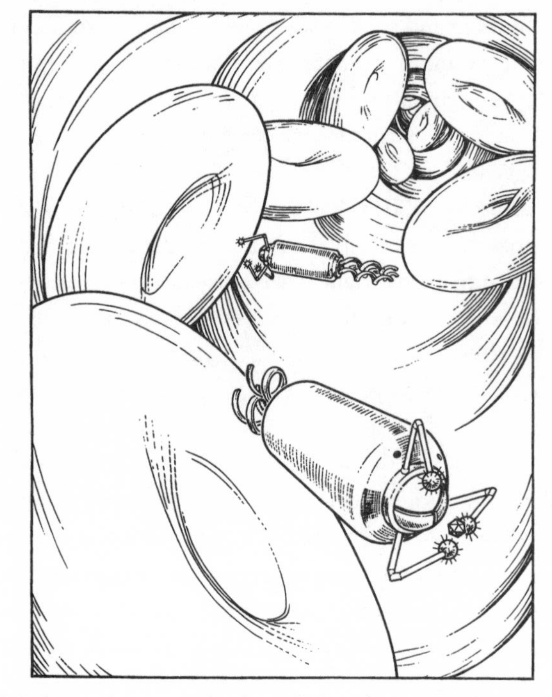

Figure 9.5 A fantasy in nanotechnology. Robot devices sent in to repair red blood corpuscles.

cult fanatics—who are now saying that something like Figure 9.5 will be a reality in the not too distant future. If they are right, there is hardly an area of human life that will not be dramatically affected. Doctoring is an example. Modern surgeons are highly skilled people, with delicate, precision instruments. To remove the lens of an eye when it is clouded with cataract and replace it with a substitute lens, as modern surgeons do, is an amazing feat of skill. The instruments that they use are impressively fine and precise. But, compared with the scale of nanotechnology, they are still immensely crude. Listen to Eric Drexler, the American scientist who is emerging as the high priest of nanotechnology, as he invokes a nano-scale view of present-day scalpels and surgical stitching.

> Modern scalpels and sutures are simply too coarse for repairing capillaries, cells, and molecules. Consider 'delicate' surgery from a cell's

perspective: a huge blade sweeps down, chopping blindly past and through the molecular machinery of a crowd of cells, slaughtering thousands. Later, a great obelisk plunges through the divided crowd, dragging a cable as wide as a freight train behind it to rope the crowd together again. From a cell's perspective, even the most delicate surgery, performed with exquisite knives and great skill, is still a butcher job. Only the ability of cells to abandon their dead, regroup, and multiply makes healing possible.

The 'obelisk', of course, is a delicate surgical needle, and the cable as wide as a goods train is the finest surgical thread. Nanotechnology holds out the dream of constructing surgical instruments small enough to work on the same scale as the cells themselves. Such instruments would be far too tiny to be controlled by the fingers of a surgeon. If a piece of thread is the width of a goods train on the cells' scale, think how wide a surgeon's fingers would be. There would have to be little automatic machines, tiny robots, not unlike miniature versions of the industrial robots we met earlier in this chapter.

Now a robot this small might be wonderful at repairing, say, a diseased red blood cell. But there is a daunting army of red blood cells for the robot to get round, about 30 billion in each one of us. So, how on earth can the little nanotechnology robot cope? You will already have guessed the answer: exponential multiplication. The hope is that the nanotechnology robot would use the same self-multiplying technique as the blood cells themselves. The robot would clone itself, replicate itself. By using the power of exponential growth, the population of robots is supposed to soar up into the billions in just the same way as the population of red blood cells went up into the billions.

Nanotechnology of this kind is all in the future, and it may never come to anything. The reason the scientists who are proposing nanotechnology think it is worth a try is this. They know that, however strange and alien it may seem to us, the equivalent of it already does work in our cells. The world of DNA and protein molecules is a world that really works on a scale that, if we did it, would be called nanotechnology. When a doctor inoculates you with immunoglobulins to stop you getting hepatitis, the doctor is loading your bloodstream with the natural equivalent of nanotechnology tools. Each immunoglobulin molecule is a complicated object which, like any other protein molecule, relies upon its shape to do its job (Figure 9.6). These little medical instruments work only because there are millions of them. They have been mass- produced—cloned up—using exponential population-growth techniques. In this case they are biological techniques: they are often grown up in the blood of a horse, for instance. Other vaccines prompt the body's own tendency to clone up antibodies like the horse immunoglobulins. The hope is that nanotechnology tools, looking pretty much like miniature industrial robots, might be cloned up too by cunningly designed artificial procedures.

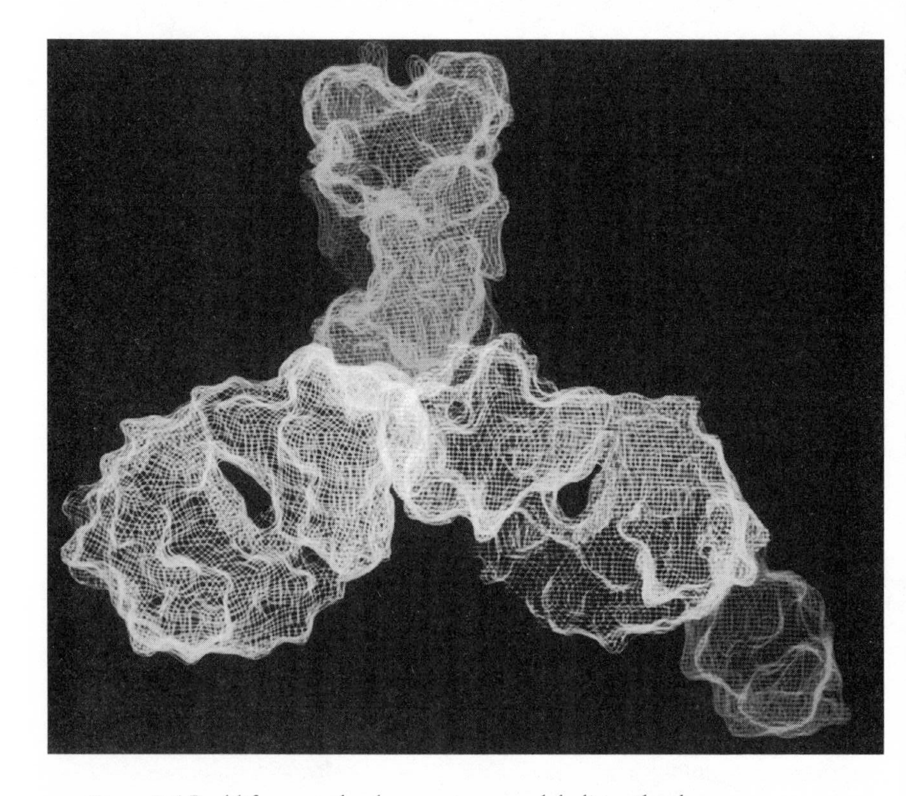

Figure 9.6 Real life nanotechnology: an immunoglobulin molecule.

Nanotechnology seems to us very strange, scarcely believable. The world of machines down there at the level of the atoms seems an alarmingly alien world, more strange than life on other planets as imagined by science-fiction writers. Nanotechnology is, for us, something that may come about in the future. It is something exciting, perhaps a bit scary and apparently new. But, far from nanotechnology being really new and alien, it is *old*. It is we big things that are new, alien, strange. We are products of a flashy new (only a few hundred million years) gigatechnology (giga from the point of view of our genes). Fundamentally, life is based in the nanoworld of the very small (nano from our point of view), a world of protein molecules, made to the coded specifications of DNA molecules and controlling the interactions of other molecules.

Nanotechnology is for the future. Let us return to the main message of this chapter and of the previous one. The genes of an elephant or a human, like the genes of a virus, can be seen as a Duplicate Me computer program. Virus genes are coded instructions that say (if they happen to be parasitizing an elephant): 'Elephant cells, duplicate me.' Elephant genes say: 'Elephant cells, work together to make a new elephant, which must be programmed in its turn to grow and make more elephants, all programmed to duplicate me.' The principle is the same. It is just that some Duplicate Me programs are more indirect and longwinded than others. Only parasitic programs can afford to be shortwinded, because they use ready-made machinery to obey their instructions. Elephant genes are not so much non-parasitic programs as mutually parasitic sub-routines. An elephant's genes are like a gigantic colony of mutually supportive viruses. Each elephant gene plays a part that is no larger than the role played by a virus gene. Each one plays its own small part in the cooperative building of the machinery that they all need for their program execution. Each one flourishes in the presence of the others. Virus genes also flourish in the presence of the set of cooperating elephant genes, but they do not contribute anything positive in return. If they did, we should probably not call them virus genes but elephant genes. To put it another way, every body contains social and antisocial genes. The antisocial ones we call virus genes (and other kinds of parasite genes). The social ones we call elephant (human, kangaroo, sycamore, etc.) genes. But the genes themselves, whether social or antisocial, whether virus genes or 'own' genes, are all just DNA instructions, and they all say, in one way or another, by fair means or foul, briefly or longwindedly, 'Duplicate me.'

CHAPTER 10

'A GARDEN INCLOSED'

WE HAVE COME A LONG WAY AND ARE FINALLY READY to return to the most difficult and complicated of all my stories, that of the fig. Let's begin with the following which sounds, at first hearing, like just another literary tease worthy of the unfortunate lecturer whom I lampooned in my opening paragraphs. A fig is not a fruit but a flower garden turned inside out. It looks like a fruit. It tastes like a fruit. It occupies a fruit-shaped niche in our mental menus and in the deep structures recognized by anthropologists. Yet it is not a fruit; it is an enclosed garden, a hanging garden and one of the wonders of the world. I am not going to leave this statement dangling as a self-indulgent profundity to be plucked by the 'sensitive' and baffle everyone else. Here is what it means.

The meaning is rooted in evolution. Figs are descended, via a chain of infinitesimally graded intermediates, from ancestors that were superficially very different from modern figs. Imagine a time-lapse film built up as follows. The first frame is a modern fig, picked today from the tree, sectioned down the middle, laid on a sheet of card and photographed. Frame two is a similar fig from a century ago. Carry on through the centuries, fig on fig, frame by frame, through a fig that might have been eaten by Jesus, or plucked by a slave for Nebuchadnezzar in the Hanging Gardens of Babylon, a fig from the land of Nod, East of Eden, figs that sweetened the short, sugar-starved lives of *Homo erectus, Homo habilis* and little Lucy of the Afar;

back before the time of cultivation, back to the wild figs of the forest and beyond. Now run the film and watch the modern fig transform itself into its remote ancestor. What changes shall we see?

Undoubtedly there'll be some shrinking as we go backwards, for cultivated figs have been plumped up over the centuries from smaller, harder, wild ancestors. But this is a superficial change and, interesting as it is, it'll all be over within the first few millennia of our backward journey. More radical and startling is the change that we'll see as we run the film further back through millions of years. The 'fruit' will open out. The tiny, almost invisible hole at the apex of the fig will pout, gape, yawn until it is no longer a hole but a cup. Look carefully at the inner surface of the cup and you'll see that it is lined with tiny flowers. First the cup is a deep one then, as we reel the film back in time, it becomes steadily shallower. Perhaps it goes through a flat stage like a sunflower, for a single sunflower too is in truth hundreds of small flowers, packed into a mass bed. Pressing on beyond the sunflower stage, our fig cup turns inside out until the florets are on the outside, as in a mulberry (the fig is a member of the mulberry family). Further back, beyond the mulberry stage, the florets separate and become more recognizably distinct flowers as in a hyacinth (although hyacinths are not closely related to figs).

Is it perhaps a little contrived, pretentious even, to describe a single fig as 'a garden inclosed'? After all, you'd hardly describe a hyacinth or a mulberry as a garden exposed. My defence is a good one and goes beyond being haunted by a phrase from the Song of Solomon. Look at a garden through the eyes of the insects that pollinate its flowers. A garden, on the human scale, is a population of flowers covering many square yards. The pollinators of figs are so tiny that, to them, the whole interior of a single fig might seem like a garden, though admittedly a small, cottage garden. It is planted with hundreds of miniature flowers, both male and female, each with its own diminutive parts. Moreover the fig really is an enclosed and largely self-sufficient world for the minuscule pollinators.

The pollinators are technically wasps belonging to one family, the Agaonidae, and they are tiny, too small to be seen clearly without a lens. By 'technically' wasps I mean that, although you might notice no great resemblance to the yellow-and-black 'yellowjackets' that menace the summer jam jar, fig wasps share a wasp ancestor with them. Fig flowers are pollinated only by these tiny wasplets (Figure 10.1). Almost every species of fig (and there are more than 900 of them) has its own private species of wasp which has been its lone genetic companion through evolutionary time since the two of them split off together from their respective predecessors. The wasps depend totally on the fig for their food and the fig depends utterly on the wasps to carry its pollen. Each species would promptly go extinct without the other. It is only the female wasps, who travel outside their natal fig, that carry the pollen. They are shaped as you might imagine highly miniaturized wasps to be. The males by contrast have no wings, for they are born and die within the closed, dark world of a single fig, and it is hard to believe that they are wasps at all, let alone wasps of the same species as their own females.

The problem with telling the life story of the fig wasps is that it is a cycle and it isn't obvious where our description should break in. This can't be helped and I'll start with the hatching of new wasp grubs, each one curled up in its tiny capsule at the base of one of the

Figure 10.1 Interior of fig with male and female fig wasps.

female flowers deep inside the enclosed garden. Feeding on the developing seed, it grows into an adult and chews its way out of its capsule, emerging into the comparative freedom of the fig's dark interior. The males and females then follow somewhat different life stories. The males hatch first and each one scours the fig searching for the capsule of an unborn female. When he finds one, he chews his way through the ovule wall and mates with the still unborn virgin. She then emerges from her birth capsule and sets off on her own journey through the miniature hanging garden. The details of what happens next vary a little from species to species. The following is typical. The female looks for male flowers, which are often to be found near the entrance to the fig. Using custom-built pollen-brushes on her front legs, she works away in the dark, systematically shovelling pollen into special pollen pockets in the recesses of her breast.

It is revealing that she takes such deliberate pains to stock up with pollen and that she has special pollen-carrying receptacles. Most pollinating insects simply find themselves dusted, willy-nilly, with the stuff. They don't have dedicated pollen-carrying apparatus, nor pollen-loading instincts. Bees do. They have pollen baskets on their legs which bulge yellow or brown with pollen stuffed into them. But bees, unlike fig wasps, are transporting the pollen to feed to their larvae. Fig wasps don't transport pollen for food. They deliberately take it on board, using special pollen-carrying pockets, for the sole purpose of fertilizing figs (which benefits the wasps only in a more indirect way). We'll return to the whole matter of the apparently amicable cooperation between figs and their pollinators.

Laden with precious pollen, the female leaves the fig for the airy openness of the outer world. Exactly how she gets out varies from species to species. In some, she crawls through the 'garden gate', the little hole at the end of the fig (Figure 10.2). In other species, it is the males' task to cut a hole through the wall of the fig through which the females leave, and they do so in collaboration, with dozens of males working together. The male's role in life is now over but the female has her big moment still before her. She flies off in the unaccustomed air, searching, probably by smell, for another fig of her own single right species. The individual fig that she seeks must also be in

Figure 10.2 The garden gate: exterior of fig to show the entrance.

the right phase of its life, the phase in which female flowers are ripe.

Having found the right kind of fig, the female locates the minute hole at the tip of it and crawls through into the dark interior. The door is so narrow that she is likely to tear her wings out by the roots as she squeezes through. Investigators who have examined fig pores have found them clogged with disembodied wings, antennae and other wasp fragments. From the fig's point of view, the advantage of having such an uncomfortably narrow entrance is that it keeps out unwanted parasites. The wing-tearing gauntlet run by the female probably also serves to wipe her clean of bacteria and harmful dirt. From the wasp's point of view, even if it is painful to have her wings ripped out by the roots, she is never going to need them again and they would probably have hampered her movements in the confines of the enclosed garden. Recall that queen ants often bite off their own wings when they have finished their mating flight and have reached the stage where wings would get in the way underground.

Inside the fig, the female wasp sets about her final mission before she dies, and it is a dual one. She pollinates all the female flowers that she visits inside the fig, and she lays eggs in some of them. Not in all. If she laid eggs in all the flowers of a fig, that fig would have failed as an organ of reproduction for the tree—all its seeds would have been eaten by wasp grubs. Does this sparing of some flowers betoken altruistic restraint on the part of the wasp? That's a question that needs careful handling. There are certain Darwinistically respectable ways in which a kind of restraint on the part of wasps could evolve into existence. But there are at least some species in which the fig tree looks after its own interests by rationing the number of flowers in which the wasp is allowed to lay eggs. The techniques are ingenious and I'll turn aside from my account of the normal life cycle just long enough to describe two of them.

In some species, the fig contains two kinds of female flowers, long-styled and short-styled. (The style is the spiky female part that sticks up in the middle of any flower.) The wasp tries to lay eggs in flowers of both kinds but her ovipositor is too short to reach the base of the long-styled flowers so she gives up and moves on. Only when she connects with a short-styled flower does her ovipositor hit bottom, and she lays one egg. In other species of fig which lack the distinction between long-styled and short-styled flowers, the fig tree's method of policing wasp behaviour can be more draconian. Or so believes W. D. Hamilton, now my Oxford colleague and one of Darwin's foremost successors today. Hamilton suggests, with some support from his own observations in Brazil, that fig trees can detect when a fig has been over-exploited by wasps. Figs that have eggs laid in all their flowers are useless from the tree's point of view. The wasps have been too selfish. They have killed the goose that lays golden eggs. Or rather, according to Hamilton, the goose commits suicide. The fig tree causes an over-exploited fig to drop to the ground where all the wasp eggs inside perish. It is tempting to see this as revenge, and there are theoretically respectable mathematical models which could free us from the suspicion of anthropomorphism. But in this case probably what the tree is doing is not so much wreaking vengeance as cutting its own losses. It costs resources to ripen a fig, and resources are wasted if used to ripen a fig ruined by greedy wasps. This kind of strategic game language, by the way, in which we are not afraid to use words like 'revenge' and 'policing', will recur throughout this chapter. It is legitimate if handled properly, which will often turn out to mean using the mathematical theory of games.

Returning to the life cycle of the typical fig wasp, our female had just wriggled like Alice through the tiny door, never to see the outside world again, and she set about relieving herself of the pollen she had gathered in the fig of her birth. The pollination behaviour of female fig wasps has a form which makes it seem deliberate. Far from letting the pollen be accidentally brushed off her body, as happens with most pollinating insects, the female of at least some species of fig wasp unloads her cargo with the same industry and attention that she lavished on loading up. She again uses the brushes on her front legs, systematically shovels pollen out of her custom-made pockets on to the brushes, and vigorously shakes it on to the receptive surface of the female flower.

By laying her eggs in female flowers the female wasp brings our story of the life cycle to its close. Her own life is over, too. She crawls to some damp crevice of the enclosed garden and dies. She dies, but she leaves behind megabits of genetic information faithfully recorded in her eggs, and the cycle resumes.

Give or take a few details which I'll enlarge on in a moment, the story I have told is similar for most kinds of figs. *Ficus,* the fig genus, is one of the largest in the living kingdoms. It is also a very diverse genus. In addition to the two species of edible (by us) figs, the genus

Figure 10.3 *(a)* strangler fig; *(b)* baobab tree entwined by strangler fig.

includes the rubber tree, the sacred banyan tree, the Bo tree *Ficus religiosa,* under which the Buddha contemplated, various shrubs and creepers, and the sinister 'strangling' figs of the tropics. The story of the strangling figs is worth telling. The forest floor is a dark place, starved of solar energy. It is the goal of every tree in the forest to reach the open sky and the sun. Tree trunks are leaf-elevators, devices for lifting solar panels—leaves—above the shade of rival trees. Most trees are fated to die as saplings. Only when an adult tree in the immediate vicinity crashes down, overcome by gales and years, does a young sapling have its chance. At any one point in the forest, this lucky event may happen just once in a hundred years. When it does, there is a gold rush to the sun. All the saplings in the area, drawn from many species, enter a headlong race to be the one to fill the precious gap in the canopy.

But the strangling figs have discovered their own sinister short cut and their story would upstage the serpent of Genesis (Figure 10.3). Instead of waiting for an existing tree to die, they contrive the event. A strangling fig tree begins life as a climber. It wraps itself around an existing tree of another species and grows like a clematis or rambling rose. But, unlike a clematis, the strangling fig's tendrils continue to grow stouter and stronger. It relentlessly tightens its grip on the unfortunate host tree, preventing it growing and eventually achieving the botanical equivalent of throttling it to death. The fig tree has by now grown to a respectable height, and it easily wins the race to the patch of light vacated by the stifled tree. The banyan tree is a kind of strangling fig with an added, remarkable, feature. Having smothered its original host, it sends out aerial roots which, when they hit ground, become proper, absorbing roots but, above ground, serve as additional trunks. So the single tree becomes an entire wood which may be 1,000 feet in diameter and can provide shelter for a medium-sized covered market in India.

I've been telling fig stories partly to show that the facts about figs are at least as enthralling as anything that my lecturer of Chapter I was able to dig up in mythology or literature, but also to illustrate a scientific way of tackling questions which might serve as a salutary example to that literary dilettante. The facts that I have so briefly recounted are the product of many man-years of meticulous and ingenious work: work that deserves the accolade 'scientific', not because it employed elaborate or expensive apparatus but because it was disciplined by a certain attitude of mind. Much of the deciphering of the wasp-pollination story would simply have involved slicing figs open and looking inside. But 'looking' gives too laid-back an impression. It wasn't a passive gawping but a carefully planned recording session yielding numbers to be fed into calculations. Don't just pluck figs and slice them. Systematically sample figs from a large number of trees, from particular heights, and at particular seasons of the year. Don't just stare at the wasps wriggling inside: identify them, photograph them, accurately draw them, count them and measure them. Classify them by species, sex, age and location in the fig. Send specimens to museums for identification by detailed comparison with internationally recognized standards. But don't make measurements and counts indiscriminately just for the sake of it. Make them in the service of testing stated hypotheses. And when you look to see if your counts and measurements fit the expectations of your hypothesis, be aware, in calculated detail, how likely it is that your results could have been obtained by chance and mean nothing.

But let's return to the fig wasps themselves. I said that, in the case of many species of fig wasps, the males in a fig collaborate to dig a hole through which all the females can escape. Why? Why doesn't a male, given that his colleagues are going to make the hole, sit back and leave them to it? Here, in microcosm, is a puzzle that continually intrigues biologists: the puzzle of altruism. An additional problem hampers the biologist seeking to explain the matter to non-specialists.

Common sense seldom perceives it as a puzzle at all. So the biologist, before he can get down to extolling the ingenuity of the solution, has to begin by persuading his audience that there ever was a puzzle requiring a special solution in the first place. In the special case of the male fig wasps, the reason it is a puzzle is this. A male that did sit back and let his colleagues make the hole would be able to save up all his energy for mating with females, secure in the knowledge that he need not hold himself back for the effort of making the hole. Other things being equal, genes for refusing to help would spread at the expense of rival genes for cooperating in hole-making. To say that genes for X will spread at the expense of Y is tantamount to saying that Y will disappear from the scene, supplanted by X. Of course, the consequence of this would be that no hole would be dug, and all the males would suffer. But this is not in itself a reason to expect males to dig. It might be a reason if they had human foresight but, assuming that they haven't, natural selection will always favour short-term benefit. Given that all the rest of the males are digging, short-term benefit will be enjoyed by an individual male that opts out and saves his energy. On this argument, digging should disappear from the population, driven out by natural selection. The fact that this doesn't happen presents us with a puzzle. Fortunately it is a puzzle that we know, in principle, how to solve.

Part of the solution may lie in kinship: in the high probability that all the males in a given fig are brothers. Brothers tend to share copies of the same genes. A wasp that helps to dig a hole will be releasing not only females with whom he has mated but females with whom his brothers have mated. Copies of the genes that foster cooperative digging will pour out through the hole, riding in the bodies of all these females. That is why those genes persist in the world, and that is a good explanation for the persistence of the behaviour among males.

But kinship is probably not the whole answer. I shan't spell it out, but there is an element of games playing which has no connection with brotherhood and this applies to the cooperation between wasp and fig. The whole story of wasps and figs is redolent of hard bargaining, of trust and betrayal, of temptation to defect policed by unconscious retaliation. We have already had a taste of this in the

Hamilton theory about over-exploited figs dropping to the ground. As so often, I must give a ritual warning that it really is all unconscious. This is superficially obvious for the fig half of the story, since no sane person thinks plants are conscious. Wasps may or may not be, but for the purposes of this chapter we are treating wasp strategy as being on the same footing as the strategy of an indubitably unconscious fig tree.

The garden inclosed is a paradise cultivated for the benefit of small insects and not surprisingly it is home to a rich and writhing Lilliputian fauna, not just the wasps whose pollination services make it all ultimately possible. Miniaturized beetle, moth and fly larvae abound, as do mites and small worms. There are predators lurking at the very gate of the garden, waiting to cash in on the rich fauna within (Figure 10.4).

The true pollinators are not the only miniature wasps that live in figs and are lumped under the general name of 'fig wasps'. There are freeloaders, distant relatives of the bona fide pollinators and parasitic upon them. Instead of entering the fig through the hole at the top,

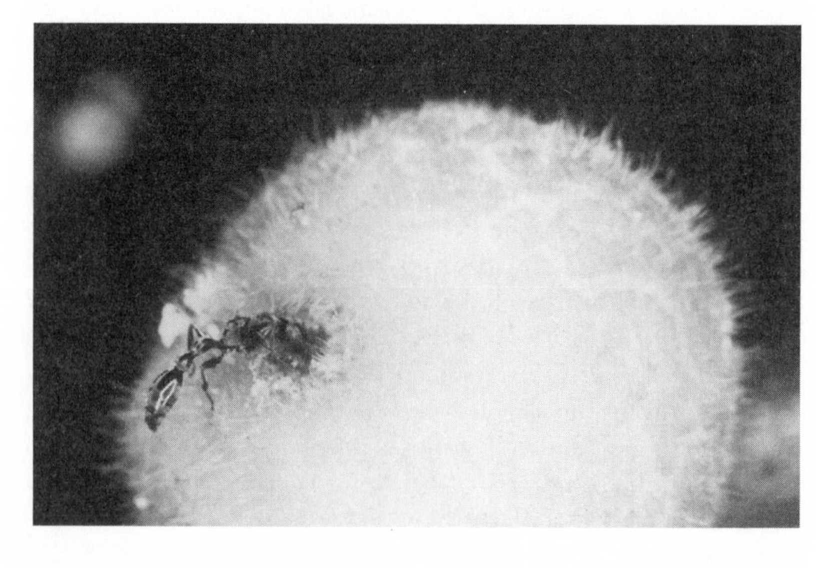

Figure 10.4 Perils of being a fig wasp. An ant lurks outside the garden gate waiting for wasps to emerge.

these parasitic wasps are normally injected as eggs through the fig wall via the spectacularly long and thin hypodermic which is their mother's uniquely specialized ovipositor (Figure 10.5). Deep inside the fig, the tip of the hypodermic seeks out the little flowers in which eggs of true, pollinating fig wasps have been laid. A female parasitic wasp looks and works like a drilling rig, and the hole that she bores through the fig wall is, on her own scale, equivalent to a 100-foot well. Males are often wingless, like the true fig wasp males (Figure 10.6). To crown the story, there are second-order parasites, wasps that lurk by the side of a 'drilling-rig' wasp, waiting for her to finish work. As soon as she pulls out, the hyperparasite slips her own, more modest ovipositor into the bore-hole and inserts her own egg.

Like the pollinating wasps themselves, the individuals of the various freeloading parasite species are playing complicated games of strategy with one another. This was investigated by the same W. D. Hamilton, working in Brazil with his wife Christine. Unlike the pollinators, the freeloader species of wasps often have winged males as well as winged females. Some species have all winged males, some spe-

Figure 10.5 Sectioned fig with parasitic wasp females waving their 'drilling rigs' in the air.

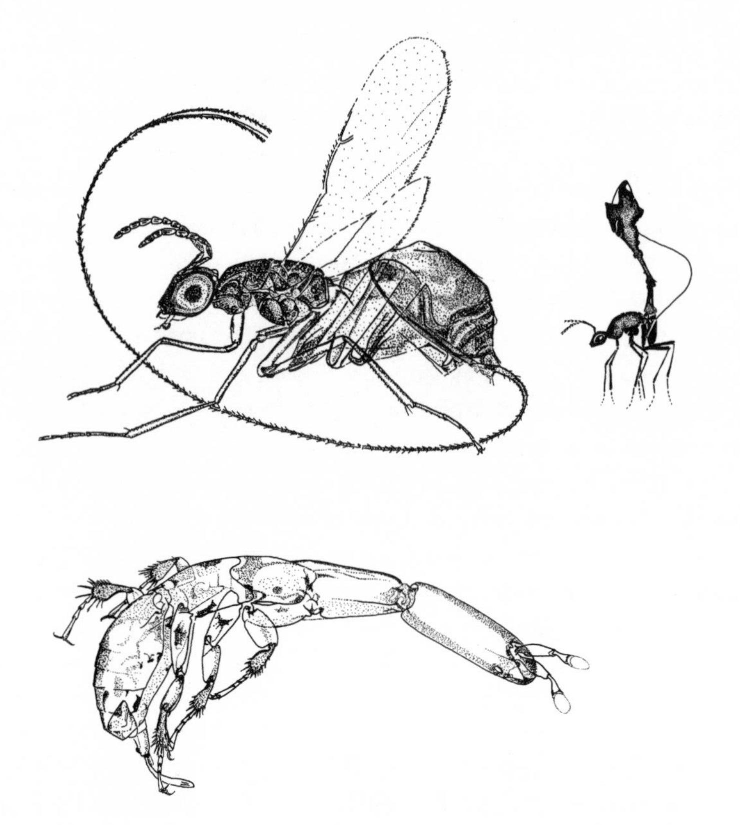

Figure 10.6 Freeloaders: parasitic wasps, *Apocrypta perplexa,* that do no pollinating but benefit from the fig. *(a)* Female; *(b)* miniature view of female in 'drilling rig' position; *(c)* male, with no wings and apparently nothing like a wasp.

cies have all wingless males and some species have a mixture of winged and wingless males. Wingless males, like the males of pollinator species, never leave their natal fig where they fight, mate and die. Winged males are like females in flying out of their natal fig, where they will mate with any females that have not already mated. So, there are two alternative ways of being a male, and some species exhibit both ways. Interestingly, the rarest species are most likely to be winged and the commonest species most likely to be wingless. This makes sense because a male of a common species is pretty likely to find a female of his own species in the same fig. A male of a rare species, however, is likely to be the only member of his species in his fig. His best hope of finding a mate is to fly away to look for one. Indeed, the Hamiltons found that winged males actually refuse to mate *until* they have flown out of their natal fig.

From a strategic point of view we are especially interested in those species that have two kinds of males. It's almost like having a third sex. In fact the winged males *look* far more like females than they look like wingless males. Both females and winged males are almost believable as wasps, although they are tiny. But the wingless males are nothing like wasps to look at. Many have savage pincer jaws which make them look a bit like miniature earwigs going backwards. They seem to use these jaws only for fighting—lacerating and slicing to death other males that they encounter as they stalk the length and breadth of the dark, moist, silent garden that is their only world. Professor Hamilton gives us a memorable description.

> Their fighting looks at once vicious and cautious—cowardly would be the word except that, on reflection, this seems unfair in a situation that can only be likened in human terms to a darkened room full of jostling people among whom, or else lurking in cupboards and recesses which open on all sides, are a dozen or so maniacal homicides armed with knives. One bite is easily fatal. One large *Idarnes* male is capable of biting another in half, but usually a lethal bite is quite a small puncture in the body. Paralysis follows a small injury so regularly and quickly as to suggest use of venom . . . If no serious injury results from the first or second reciprocal attempts to bite, one of the males, injured perhaps by loss of a [foot] or in some way sensing himself outmatched, retreats and tries to hide . . . From this position he can bite at the legs of the victor or another passing male with much less danger . . . One fruiting of a large tree of *Ficus* probably involves several million deaths due to combat.

The phenomenon of a species having two distinct kinds of males is not unknown elsewhere among animals, but it is never so pronounced as in the case of the fig freeloader wasps. There are individual red deer stags called hummels which lack antlers yet seem to make a respectable showing of reproducing themselves in competition with their antlered rivals. Theorists have distinguished two possibilities for what is going on in such cases. One is the 'best of a bad job' theory. This probably applies to a species of solitary bee called *Centris pallida*. The two kinds of male bee are called 'patrollers' and 'hoverers'. Patrollers are large. They actively search for females who have not yet hatched out of their underground nurseries, dig down and mate with them underground. Hoverers are small. They don't dig but hover in the air, waiting for the emergence of the few flying females that the patrollers missed underground. Evidence suggests that the patrollers do better than the hoverers but, given that you are a small male with little chance of succeeding as a patroller, you can make the best of a bad job by hovering instead. As always, this is genetic, not conscious, choice.

The other theory about how two kinds of males can co-exist in a species is the stable balance theory. It seems that this may apply in the case of the fig freeloaders. The idea here is that both kinds of male succeed equally well when they exist in a special, balanced proportion in the population. What keeps the proportion balanced is this. When a male is a member of the rarer type he does well, specifically because he is rare. Therefore more of his kind are born, and they consequently cease to be rare. If they succeed so well that they become common, the other kind now have the advantage by virtue of being relatively rare and they accordingly become commoner again. So the proportion is regulated rather in the manner of a thermostat. I've told the story as if it gave rise to wild oscillation but this there need not be, any more than a thermostatically controlled room oscillates wildly in temperature. Nor does the stable equilibrium proportion have to be 50–50. Whatever the equilibrium proportion may be, natural selection keeps pushing the population back towards it. The equilibrium proportion is that proportion at which the two types of male do equally well.

How might something like this work itself out in the case of the fig freeloaders? The first fact we need is that the females of these parasitic species tend to lay only one or two eggs in a fig before moving on to another one (they poke their ovipositor in, you'll remember,

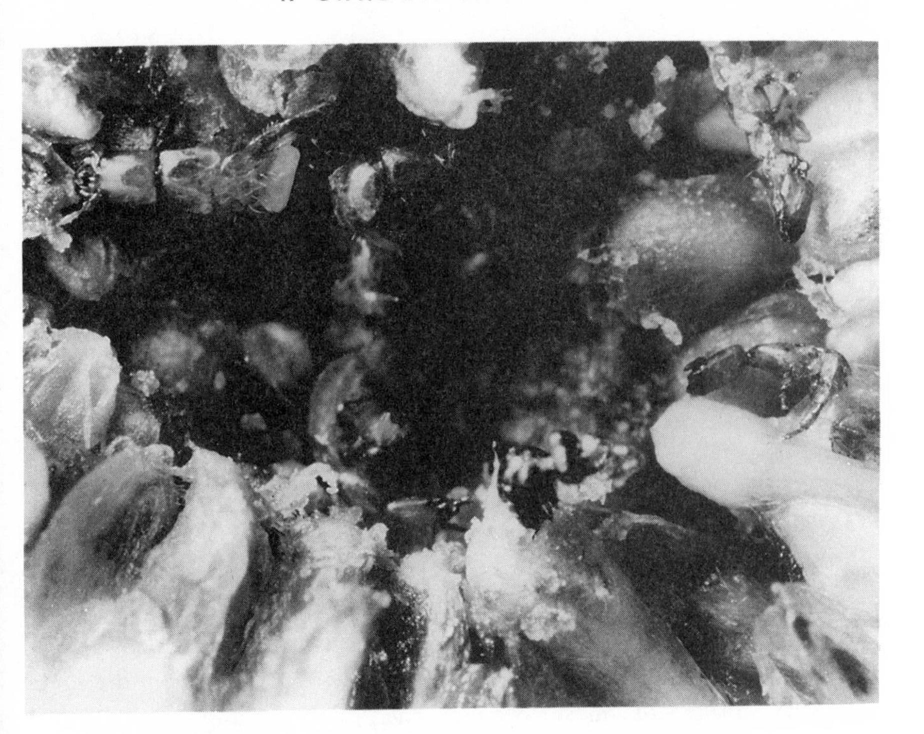

Figure 10.7 A Garden Inclosed.

from the outside of the fig). There are good reasons for this. If a female placed all her eggs in one fig, her daughters and (wingless) sons would be likely to mate with each other, and it is well understood that incest is a bad thing, for the same kind of reason as flowers avoid self-pollination. Anyway, it is a fact that the females do spread their progeny around thinly among figs. A consequence is that there will, by luck, be a number of figs that happen not to have any eggs of the species at all. And there'll be a number that happen to have no male eggs and a number that happen to have no female eggs.

Now think about the possibilities that might face a wingless male wasp. If he hatches out in a fig with no females, there is nothing he can do. That is the end of him, genetically speaking. But if there are any females at all in his fig, he has a good chance of mating with them, albeit in competition with other males of his kind—and it is no wonder these tiny male wasps are among the best armed and most ruthless fighters in the animal kingdom. Few females leave their figs unmated if there are any wingless males inside. Some figs will happen to have female eggs but no male eggs. These females will leave the fig unmated, and the only males that can mate with them are winged males outside.

Winged males, then, can prosper only if there are some figs that happen to have female wasps but no wingless males. How likely is this? It depends upon how densely populated the wasps are, relative to the figs. And it also depends upon the fraction of males that have wings. If the wasps in general are rare compared to figs, wasp eggs will be few and far between, and you can hope to find at least some figs that have females only. Winged males will do relatively well under these conditions. Now look what happens if the wasp population is high. Most of the figs will have several wasps in them, of both sexes. Most of the females will be mated by wingless males before they leave their figs, and the winged males will do badly.

Hamilton did these calculations more precisely. He concluded that if the average number of male eggs per fig is greater than about three, winged males will almost never reproduce. At all densities higher than this, selection will favour winglessness among the males. If the average number of male eggs per fig is one or fewer, wingless males will do badly because they'll almost never find themselves sharing a fig with another wasp, let alone a female. Natural selection under these conditions will favour wings among males. At intermediate population densities, the stable balance theory comes into play and natural selection favours a mixture of winged and wingless males in the population.

Once we have the stable balance theory in action, natural selection favours whichever male type is in the minority, or rather, whichever male type exists at less than the critical frequency, whatever the critical frequency is. We can then say as a shorthand that natural selection is favouring the critical frequency. As for the critical frequency itself, it will vary from species to species, depending upon the absolute density of wasps compared to figs. We can think of different species of wasp, with their various wasp/fig densities, as like different thermostatically controlled rooms. Each room has its thermostat set to a different temperature. For instance in a species where the average number of male eggs per fig is three, natural selection favours a mixture of males in the population with about 90 per cent of them wingless. In a species where the average number of males per fig is two, natural selection favours a mixture of males with about 80 per cent of them wingless. Remember that the condition we are talking about is an *average* of two males per fig. This does not mean that every fig has exactly two males. This average of two is made up of some figs with no males, some with one, some with two and some with more than two. The 20 per cent winged males make their genetic living not from the figs with two males (whose females are likely to be snapped up before they leave the fig) but from the figs with no males at all.

What is the evidence, in practice, that the stable balance theory is working among these wasps, rather than the best of a bad job theory? The key difference between the two theories is that on the first theory, but not the second, both kinds of male wasp should do equally well. The Hamiltons found evidence suggesting that the two kinds of male really do succeed equally well in mating with females. They looked at ten different species. They found that, in all species, the proportion of winged to wingless males was approximately equal to the proportion of females that left their natal figs unmated. Thus a species in which 80 per cent of the females flew out of their natal figs unmated, was also a species in which 80 per cent of the males were winged. A species in which 70 per cent of the females were mated before leaving their natal figs was also a species in which 70 per cent of the males were wingless. It really does look as though the proportions of the two kinds of males are exactly what they should be to ensure an equal share of females all round. This is evidence for the stable balance theory and against the best of a bad job theory. Sorry it was so complicated, but that is par for the course in the world of figs.

Let's leave the freeloader wasps and return to the true fig wasps, the bona fide, specialist fig pollinators. If you think the freeloader saga was complicated, get ready for this, my final story. I take some pride in trying my hardest to explain difficult and complicated matters but the following may defeat me. Let me do my best and, if I fail, blame the figs and their partners the wasps. Or, don't blame them or me but credit evolution for the subtle wonder of this complex dance through evolutionary time. It takes some work to follow these final pages of the book but I like to hope that it is worth the effort.

The figs in which the following story is set are 'dioecious'. This means that, instead of having one kind of fig containing both male and female flowers like the 'monoecious' trees we've so far talked about, there are male trees and female trees. Female trees produce figs containing only female flowers. Male trees make figs containing male flowers, but that isn't all. These so-called male figs also contain pseudo-female flowers and this is very important for the wasps. Unlike the real female flowers in the female figs, the pseudo-female flowers in the male figs cannot set seed, even if they are pollinated. What they are good at is providing food for baby wasps, and—thereby will hang a tale—they need to be pollinated before they will do this. The fertile female flowers in the all-female figs are a genetic graveyard for wasps though they are vital for fig reproduction. Female wasps enter them and pollinate them, but their eggs cannot grow in them.

Here we have the elements of a rich strategic game which we can describe in terms of the 'wants' (in the special Darwinian sense) of the various players. Both male and female figs 'want' wasps to enter them, but the wasps want to enter only male figs—and then only because of the nutritious pseudo-female flowers within them. The male tree wants eggs to be laid in its pseudo-female flowers so that the female young that hatch out will then load themselves up with the male fig's pollen and fly off with it. The tree has no direct interest in male wasps hatching inside its figs because male wasps don't transport pollen. This may seem surprising since male wasps are, after all, necessary for the continuation of the race of fig wasps. We people, with our penchant for looking ahead and thinking of the larger consequences of our actions, find it hard to purge our minds of the idea that natural selection too looks ahead. I have already made this point in another connection. If natural selection were capable of taking the long view, animals and plants would take steps to preserve the race—their own and those that they depend on like their prey and their pollinators. But nature, unlike humans with brains, has no foresight. 'Selfish genes' and short-term benefit are always favoured in a world where others are coping with the long-term needs of the race. If an individual fig tree could get away with fostering nothing but female wasps it would do so, relying on other fig trees to produce the males needed to preserve the race of wasps. The point is that, as long as the other trees are producing male wasps, a single selfish rebel tree that discovered a way to increase its production of female wasps, and hence increase the amount of pollen that it could export, would have an advantage. As the generations went by, more and more trees would become selfish, relying on fewer and fewer trees to produce the needed male wasps. Finally, the last tree with a penchant for nurturing male wasps would die because it would still be doing less well than its rivals producing only female wasps.

Fortunately, it appears that fig trees have no control over the sex ratio of the wasps reared in their figs. If they could control it, it is likely that male figs would disappear and the race of fig wasps would die out. The fact that the race of fig trees would then also die out would be just too bad. Natural selection cannot look that far ahead. The reason fig trees have no control over wasp sex ratios is probably that the wasps, who also have an interest in controlling their own sex ratio, have overriding power.

A female fig tree also wants (again in the special Darwinian sense) female wasps to enter its figs, otherwise its female flowers won't be pollinated. A female wasp wants to enter male figs, because only there will she find the pseudo-female flowers in which her larvae can grow. She wants to avoid female figs like the plague because once she has entered one she is genetically dead. She will have no descendants. Expressed more strictly, genes that make wasps enter female figs will not be passed on to future generations. If natural selection were working on wasps alone, the world would become full of wasps that discriminate against female figs and in favour of male figs with their lovely pseudo-female egg-cosies.

Once again, we humans want to interrupt and say: 'But surely the fig wasps should want some of their number to enter female figs because, though these may be graveyards for individual wasp genes, they are vitally important for the continuation of the race of fig trees. If the race of fig trees goes extinct, the race of fig wasps would soon follow them.' This is the exact mirror image of our previous argument. Given that some fig wasps are foolish enough, or altruistic enough, to enter female figs, natural selection will favour a selfish individual wasp that discovers how to avoid female figs and enter only male figs. Selfishness among wasps is bound to be favoured over any public-spirited tendency to work for the continuation of the race. So why don't fig trees and fig wasps go extinct? Not because of altruism or foresight, but because selfishness on each side of the wasp/fig divide is prevented by selfish counter-measures on the other side. What prevents female wasps from doing the selfish thing and avoiding female figs is direct action taken by the fig trees themselves to thwart would-be selfish wasps. Natural selection has favoured deceptive tactics by female figs, making them become so like male figs that the wasps can't tell the difference.

So, our game between wasps and figs has a fascinating symmetry. There are opportunities on both sides for individuals to be selfish. If either of these two selfish impulses succeeded, both the wasps and the fig trees might go extinct. What stops this happening is not altruistic restraint, nor ecologically aware foresight. What stops it happening is direct police action by individual players on the other side in each case, acting in their own selfish interest. Fig trees would, if they could, abolish male wasps, incidentally thereby ensuring the wasps', and their own, extinction. They are prevented from doing so by the wasps, who have an interest in rearing both male and female wasps. Fig wasps would, if they could, avoid entering female figs and incidentally thereby ensure the trees' and their own extinction. They are prevented from doing so by the trees, who make it difficult to tell the difference between male and female figs.

To summarize so far, we can expect both male and female fig trees to do all in their power to lure wasps into their own kind of fig. And we can expect the wasps to struggle to tell the difference between male and female figs, to enter the first and shun the second. Remember that 'struggle' means that, over evolutionary time, they will come to possess genes that confer a predilection for male figs. More contortedly, we shall also find that both male and female fig trees should have an interest in fostering wasps that enter figs of the *other* sex. In this difficult argument, I am following a brilliant paper by two British biologists: Alan Grafen, one of modern Darwinism's foremost mathematical theorists, and Charles Godfray, a leading ecologist and entomologist.

What weapons do fig trees have in playing their game of strategy? Female trees can make their figs look, and smell, as much like male figs as possible. Mimicry, as we saw in earlier chapters, is a common phenomenon in the living kingdoms. Stick insects resemble inedible sticks and are therefore ignored by birds. Many palatable butterflies resemble distasteful butterflies of a completely different species that birds have learned to avoid. Orchids of various species mimic bees, flies or wasps. Mimicry of this kind has delighted naturalists since the nineteenth century and has often fooled collectors just as effectively as it presumably fools other animals. Although the object of awe and incomprehension in the past, it is now clear that mimicry, of almost limitless perfection, readily evolves by natural selection. Mimicry of (desirable to wasps) male figs by (undesirable to wasps) female figs is certainly to be expected, but the sequel is—to put it mildly—less obvious and needs a lot more thinking out. We also expect male figs to go out of their way to look, and smell, like female figs. Here's why.

A male tree 'wants' female wasps to enter its figs and lay eggs in the pseudo-female flowers there. But the fig gains from this only insofar as the young female wasps that subsequently hatch out go on to play their allotted role. The new females must load themselves up with pollen, leave the natal fig, and then at least some of them must enter the genetic graveyard of a female fig and pollinate it (thereby propagating the fig's genes, though not the wasps' own). A male fig that looks very unlike female figs may be very successful in helping female wasps to achieve their goal of entering only male flowers and laying their eggs. But the daughters of those wasps will tend to inherit their mothers' taste in figs. The daughter wasps will inherit a tendency to go only for male figs and they will be useless at propagating the genes of the fig in which they hatched (though good at propagating their own genes).

Now consider a rival male tree whose figs resemble female figs. It may be harder for it to lure female wasps, who will be put off since they are trying to avoid female figs. But those female wasps that it *does* manage to lure will be a specially selected subset of female wasps: they will be female wasps who have been foolish enough (from their own point of view) to enter a fig that looks like a female fig. These wasps will lay their eggs in the pseudo-female flowers, as before. As before, their daughters will inherit their mothers' taste in figs. And now, think about what that taste will be. These young wasps came from mothers who willingly and eagerly entered a male fig *that looked like a female fig,* and their daughters will inherit their (foolish from their point of view) proclivity. Their daughters will go out into the world, looking for figs that look like female figs. And a healthy proportion will find them—thereby killing their own genes but placing the male fig's pollen exactly where it wants it. These duped females throw away their own genes but they carry successful fig genes in their pollen baskets, including genes for making male figs mimic female figs. Fig genes from rival trees, genes that make male figs very different from female figs, will also be carried in wasp pollen baskets. But those basketfuls of pollen are more likely to be thrown away—from the male fig's point of view—in the genetic graveyard of other male figs. Hence male figs will 'conspire' with female figs to make it difficult for wasps to discriminate between them and avoid *their* genetic graveyard. Male figs and female figs will 'agree' in 'wanting' to be indistinguishable.

As Einstein once exulted, subtle is the Lord! But, if you can bear it, the plot thickens. The pseudo-female flowers, inside male figs, require to be pollinated if they are to provide the food that a wasp larva needs. So there is no difficulty in understanding, from a female wasp's point of view, why she actively loads herself up with pollen; no difficulty in understanding why the females have special pollen-carrying baskets rather than just becoming accidentally dusted with pollen. Female wasps have everything to gain from carrying pollen. They need pollen to provoke pseudo-female flowers into making food for their grubs. But Grafen and Godfray point out that we do still have a problem on the other side of this remarkable relationship. We have a problem when we turn back to the figs. Why do the pseudo-female flowers in a male fig *need* to be pollinated before they will nourish wasp larvae? Wouldn't it be simpler just to provide food for wasp larvae whether pollinated or not? Male figs need to feed wasp larvae so that they will take pollen away to female figs. But why do the pseudo-female flowers insist on being pollinated before they will yield food?

Imagine a mutant male tree that became less fussy: a sport that relaxed this requirement and allowed wasp larvae to develop even if laid in unpollinated flowers. This mutant tree would seem to have an advantage over its more fussy rivals because it would produce a larger crop of young wasps. Think about it. Any fig will be entered by some females who, for one reason or another, don't have any pollen in their baskets. In the fussy fig, these females may lay eggs but the resulting larvae will starve and no young pollinators will result. But now look at the rival, mutant, unfussy fig. If it is entered by a pollen-less female, no matter. Her larvae will grow regardless and will hatch into healthy young wasps. The unfussy fig will produce a larger crop of young wasps because it will rear the progeny not only of pollen-carrying wasps but also of wasps that failed to carry pollen. So the unfussy male fig will have a clear advantage over the fussy male fig, because it will produce a larger army of young female wasps to carry its pollen off and into the genetic future. Won't it?

No it won't, and here is the almost too convoluted subtlety that Grafen and Godfray discerned. This great little army of young female wasps, swarming out of the unfussy fig, will indeed be numerous. But—the argument is like the previous one—they will inherit the proclivities of their mothers. Their mothers—specifically the mothers of the surplus wasps, the wasps that the unfussy fig produces over and above its fussy rivals—had a flaw. They failed to pick up pollen, or for some other reason failed to pollinate the flower in which their larvae grew. This is why the extra larvae are extra at all. And the extra larvae will tend to inherit the flaw. They will tend not to pick up pollen, or otherwise will tend to be bad pollinators. It is almost as though the fussy male fig deliberately imposes a hurdle for the wasps that enter it. It tests them to see if they will do to the pseudo-female flowers everything that they would have to do to a real female flower. If they do not, their larvae are not allowed to develop. By imposing this test, the male fig is selecting those wasp genes that tend to make wasps good at passing on fig genes. Grafen and Godfray call it 'vicarious selection'. It is a little bit like artificial selection, such as we met in Chapter I, and yet not completely like it. Pseudo-female flowers are like simulators used to weed out pilots unqualified to fly real planes.

Vicarious selection is a novel idea and it provides the answers to even more subtle problems. Fig genes and wasp genes are partners, locked together in a fast waltz through geological time. Most of the many species of fig have, as we've seen, their own private species of wasp. Figs and their wasps have evolved together—'*co*-evolved'—in step with each other and out of step with other fig and wasp species. We have seen the advantage of this from the figs' point of view. Their private species of pollinating wasp is the ultimate magic bullet. By cultivating one, and only one, species of wasp, they target their pollen strictly to female figs of their own species and no other. They do not waste pollen the way they would if they had to share the same species of wasp, one species that promiscuously visits all fig species. Whether such strict loyalty to one fig species also benefits the wasps is less clear, but they probably have no choice. For reasons that we need not go into, species occasionally evolve away from one another, splitting into two species. In the case of fig trees, when they diverge in evolutionary time they may well change the chemical passwords by which wasps recognize figs, and perhaps also such lock-and-key details as the depth of their tiny flowers. Wasp species are forced to follow suit. For instance, gradually deepening flowers on the fig (lock) side of the coevolution impose gradually lengthening ovipositors on the wasp (key) side of the co-evolution.

Now comes a peculiar problem recognized by Grafen and Godfray. Let's expand the lock-and-key analogy. Fig species evolve away from each other by changing their locks, and wasps follow suit with their keys. Something like this must have gone on when ancestral orchids diverged into bee orchids, fly orchids and wasp orchids. But there it is easy to see how the coevolution took place. Figs raise a very special and very tantalizing problem, and it is the last problem I shall tackle in this book. If the story went according to the usual co-evolutionary plan, we should expect to see something like the following. Genes for deeper flowers, say, would be selected among the female figs. This would set up a selection pressure in favour of longer ovipositors among wasps. But because of the odd circumstances of these figs, this normal story of co-evolution can't work. The only female flowers that pass on genes are the true female ones in female figs, not the pseudo-female florets in male figs; while the only female wasps that pass on genes are the ones that lay eggs in pseudo-female flowers, not the ones that lay eggs in real female flowers. So those individual wasps that happen to have long ovipositors and succeed in reaching the bottom of the long female flowers won't pass on the genes for long ovipositors. Those individual wasps whose long ovipositors reach the bottom of the pseudo-female flowers will pass their own genes on. But here the genes for making long flowers won't be passed on. We have a riddle.

Once again, the answer seems to lie in vicarious selection—accurate simulators for pilots. Male figs 'want' the wasps that they export to be good at pollinating true female flowers. Therefore, in our hypothetical example, they would want them to have long ovipositors. The best way for a male fig to ensure this is to allow only mothers with long ovipositors to lay eggs in their pseudo-female flowers. Expressing the idea in terms of this particular example runs the risk of making it sound too purposeful, as though the male figs 'know' that female flowers are deep. Natural selection would do it automatically by favouring those male figs whose pseudo-female flowers resembled true female flowers in all respects, including depth.

Figs and fig wasps occupy the high ground of evolutionary achievement: a spectacular pinnacle of Mount Improbable. Their relationship is almost ludicrously tortuous and subtle. It cries out for interpretation in the language of deliberate, conscious, Machiavellian calculation. Yet it is achieved in the complete absence of any kind of deliberation, without brain power or intelligence of any kind. The point is rubbed home for us by the very fact that the players are a tiny wasp with a very tiny brain on the one hand, and a tree with no brain at all on the other. It is all the product of an unconscious Darwinian fine-tuning, whose intricate perfection we should not believe if it were not before our eyes. There is a form of calculation going on, or rather millions of parallel calculations, of costs and benefits. The calculations are of a complexity to tax our largest computers. Yet the 'computer' that is performing them is not made of electronic components, not even made of neural components. It is not located in a particular place in space at all. It is an automatic, distributed computer whose data bits are stored in DNA code, spread over millions of individual bodies, shuttling from body to body, via the processes of reproduction.

The famous Oxford physiologist Sir Charles Sherrington compared the brain to an enchanted loom in a famous passage:

> It is as if the Milky Way entered upon some cosmic dance. Swiftly the brain becomes an enchanted loom where millions of flashing shuttles weave a dissolving pattern, always a meaningful pattern though never an abiding one; a shifting harmony of subpatterns.

It was the rise of nervous systems and brains that brought designed objects into the world. Nervous systems themselves, and all designoid objects, are the products of an older and a slower cosmic dance. Sherrington's vision helped him to become one of the leading investigators of the nervous system in the first part of this century. We may profit by borrowing a parallel vision. Evolution is an enchanted loom of shuttling DNA codes, whose evanescent patterns, as they dance their partners through geological deep time, weave a massive database of ancestral wisdom, a digitally coded description of ancestral worlds and what it took to survive in them.

But that is a train of thought that must wait for another book. The main lesson of this book is that the evolutionary high ground cannot be approached hastily. Even the most difficult problems can be solved, and even the most precipitous heights can be scaled, if only a slow, gradual, step-by-step pathway can be found. Mount Improbable cannot be assaulted. Gradually, if not always slowly, it must be climbed.

BIBLIOGRAPHY

BOOKS REFERRED TO AND SUGGESTIONS FOR FURTHER READING

Adams, D. (1989) *The More than Complete Hitchhiker's Guide.* New York: Wings Books.

Attenborough, D. (1979) *Life on Earth.* London: Collins.

Attenborough, D. (1984) *The Living Planet.* London: Collins/BBC Books.

Attenborough, D. (1995) *The Private Life of Plants.* London: BBC Books.

Basalla, G. (1988) *The Evolution of Technology.* Cambridge: Cambridge University Press.

Berry, R. J., and Hallam, A. (Eds.) (1986) *Collins Encyclopedia of Animal Evolution.* London: Collins.

Bonner, J. T. (1988) *The Evolution of Complexity.* Princeton, NJ: Princeton University Press.

Bristowe, W. S. (1958) *The World of Spiders.* London: Collins.

Brusca, R. C., and Brusca, G. J. (1990) *Invertebrates.* Sunderland, Mass.: Sinauer.

Carroll, S. B. (1995) 'Homeotic genes and the evolution of arthropods and chordates'. *Nature,* 376, 479–85.

Coveney, P., and Highfield, R. (1995) *Frontiers of Complexity.* London: Faber and Faber.

Cringely, R. X. (1992) *Accidental Empires.* London: Viking.

Cronin, H. (1991) *The Ant and the Peacock.* Cambridge: Cambridge University Press.

Dance, S. P. (1992) *Shells.* London: Dorling Kindersley.

Darwin, C. (1859) *The Origin of Species.* Harmondsworth (1968): Penguin.

Darwin, C. (1882) *The Various Contrivances by Which Orchids are Fertilised by Insects.* London: John Murray.

Dawkins, R. (1982) *The Extended Phenotype.* Oxford: W. H. Freeman.

Dawkins, R. (1986) *The Blind Watchmaker.* Harlow: Longman.

Dawkins, R. (1989) 'The evolution of evolvability'. In *Artificial Life.* (Ed. C. Langton.) Santa Fe: Addison-Wesley.

Dawkins, R. (1989) *The Selfish Gene.* (2nd edn) Oxford: Oxford University Press.

Dawkins, R. (1995) *River Out of Eden.* London: Weidenfeld and Nicolson.

Dennett, D. C. (1995) *Darwin's Dangerous Idea.* New York: Simon and Schuster.

Douglas-Hamilton, I. and O. (1992) *Battle for the Elephants.* London: Doubleday.

Drexler, K. E. (1986) *Engines of Creation.* New York: Anchor Press/ Doubleday.

Eberhard, W. G. (1985) *Sexual Selection and Animal Genitalia.* Cambridge, Mass.: Harvard University Press.

Eldredge, N. (1995) *Reinventing Darwin: The great debate at the high table of evolutionary theory.* New York: John Wiley.

Fisher, R. A. (1958) *The Genetical Theory of Natural Selection.* New York: Dover.

Ford, E. B. (1975) *Ecological Genetics.* London: Chapman and Hall.

Frisch, K. v. (1975) *Animal Architecture.* London: Butterworth.

Fuchs, P., and Krink, T. (1994) 'Modellierung als Mittel zur Analyse raumlichen Orientierungsverhaltens'. Diplomarbeit, Universität Hamburg.

Goodwin, B. (1994) *How the Leopard Changed its Spots.* London: Weidenfeld and Nicolson.

Gould, J. L., and Gould, C. G. (1988) *The Honey Bee.* New York: Scientific American Library.

Gould, S. J. (1983) *Hen's Teeth and Horse's Toes.* New York: W. W. Norton.

Grafen, A., and Godfray, H. C. J. (1991) 'Vicarious selection explains some paradoxes in dioecious fig-pollinator systems'. *Proceedings of the Royal Society of London, B.,* 245, 73–6.

Gribbin, J., and Gribbin, M. (1993) *Being Human.* London: J. M. Dent.

Haeckel, E. (1974) *Art Forms in Nature.* New York: Dover.

Haldane, J. B. S. (1985) *On Being the Right Size.* (Ed. J. Maynard Smith.) Oxford: Oxford University Press.

Halder, G., Callaerts, P., and Gehring, W. J. (1995) 'Induction of ectopic eyes by targeted expression of the *eyeless* gene in *Drosophila*'. *Science,* 267, 1788–92.

Hamilton, W. D. (1996) *Narrow Roads of Gene Land: The collected papers of W. D. Hamilton, Vol. 1. Evolution of Social Behaviour.* Oxford: W. H. Freeman/Spektrum.

Hansell, M. H. (1984) *Animal Architecture and Building Behaviour.* London: Longman.

Hayes, B. (1995) 'Space-time on a seashell'. *American Scientist,* 83, 214–18.

Heinrich, B. (1979) *Bumblebee Economics.* Cambridge, Mass.: Harvard University Press.

Hölldobler, B., and Wilson, E. O. (1990) *The Ants.* Berlin: Springer-Verlag.

Hoyle, F. (1981) *Evolution From Space.* London: J. M. Dent.

Janzen, D. (1979) 'How to be a fig'. *Annual Review of Ecology and Systematics,* 10, 13–51.

Kauffman, S. (1995) *At Home in the Universe.* Harmondsworth: Viking.

Kettlewell, H. B. D. (1973) *The Evolution of Melanism.* Oxford: Oxford University Press.

Kingdon, J. (1993) *Self-made Man and His Undoing.* London: Simon and Schuster.

Kingsolver, J. G., and Koehl, M. A. R. (1985) 'Aerodynamics, thermoregulation, and the evolution of insect wings: differential scaling and evolutionary change'. *Evolution,* 39, 488–504.

Land, M. F. (1980) 'Optics and vision in invertebrates'. In *Handbook of Sensory Physiology.* (Ed. H. Autrum.) VII/6B, 471–592. Berlin: Springer-Verlag.

Langton, C. G. (Ed.) (1989) *Artificial Life.* New York: Addison-Wesley.

Lawrence, P. A. (1992) *The Making of a Fly.* London: Blackwell Scientific Publications.

Leakey, R. (1994) *The Origin of Humankind.* London: Weidenfeld and Nicolson.

Lundell, A. (1989) *Virus! The secret world of computer invaders that breed and destroy.* Chicago: Contemporary Books.

Macdonald, D. (Ed.) (1984) *The Encyclopedia of Mammals.* (2 vols.) London: Allen and Unwin.

Margulis, L. (1981) *Symbiosis in Cell Evolution.* San Francisco: W. H. Freeman.

Maynard Smith, J. (1988) *Did Darwin Get it Right?* Harmondsworth: Penguin Books.

Maynard Smith, J. (1993) *The Theory of Evolution*. Cambridge: Cambridge University Press.

Maynard Smith, J., and Szathm·ry, E. (1995) *The Major Transitions in Evolution*. Oxford: Freeman/Spektrum.

Meeuse, B., and Morris, S. (1984) *The Sex Life of Plants*. London: Faber and Faber.

Meinhardt, H. (1995) *The Algorithmic Beauty of Sea Shells*. Berlin: Springer-Verlag.

Moore, R. C., Lalicker, C. G., and Fischer, A. G. (1952) *Invertebrate Fossils*. New York: McGrawHill.

Nesse, R., and Williams, G. C. (1995) *Evolution and Healing: The New Science of Darwinian Medicine*. London: Weidenfeld and Nicolson. Also published as *Why We Get Sick* by Random House, New York.

Nilsson, D. -E. (1989) 'Vision, optics and evolution'. *Bioscience*, 39, 298–307.

Nilsson, D.-E. (1989) 'Optics and evolution of the compound eye'. In *Facets of Vision*. (Eds. D. G. Stavenga and R. C. Hardie.) Berlin: Springer-Verlag.

Nilsson, D.-E., and Pelger, S. (1994) 'A pessimistic estimate of the time required for an eye to evolve'. *Proceedings of the Royal Society of London, B*, 256, 53–8.

Orgel, L. E. (1973) *The Origins of Life*. London: Chapman and Hall.

Pennycuick, C. J. (1972) *Animal Flight*. London: Edward Arnold.

Pennycuick, C. J. (1992) *Newton Rules Biology*. Oxford: Oxford University Press.

Pinker, S. (1994) *The Language Instinct*. Harmondsworth: Viking.

Provine, W. B. (1986) *Sewall Wright and Evolutionary Biology*. Chicago: Chicago University Press.

Raff, R. A., and Kaufman, T. C. (1983) *Embryos, Genes and Evolution*. New York: Macmillan.

Raup, D. M. (1966) 'Geometric analysis of shell coiling: general problems'. *Journal of Paleontology*, 40, 1178–90.

Raup, D. M. (1967) 'Geometric analysis of shell coiling: coiling in ammonoids'. *Journal of Paleontology*, 41, 43–65.

Ridley, Mark (1993) *Evolution*. Oxford: Blackwell Scientific Publications.

Ridley, Matt (1993) *The Red Queen: Sex and the evolution of human nature*. Harmondsworth: Viking.

Robinson, M. H. (1991) 'Niko Tinbergen, comparative studies and evolution'. In *The Tinbergen Legacy*. (Eds. M. S. Dawkins, T. R. Halliday, and R. Dawkins.) London: Chapman and Hall.

Ruse, M. (1982) *Darwinism Defended.* Reading, Mass.: Addison-Wesley.

Sagan, C., and Druyan, A. (1992) *Shadows of Forgotten Ancestors.* New York: Random House.

Salvini-Plawen, L. v. and Mayr, E. (1977) 'On the evolution of photoreceptors and eyes'. In *Evolutionary Biology.* (Eds. M. K. Hecht, W. C. Steere, and B. Wallace.) 10, 207–63. New York: Plenum.

Terzopoulos, D., Tu, X., and Grzeszczuk, R. (1995) 'Artificial fishes: autonomous locomotion, perception, behavior, and learning in a simulated physical world'. *Artificial Life,* 1, 327–51.

Thomas, K. (1983) *Man and the Natural World: Changing Attitudes in England 1500–1800.* Harmondsworth: Penguin Books.

Thompson, D'A. (1942) *On Growth and Form.* Cambridge: Cambridge University Press.

Trivers, R. L. (1985) *Social Evolution.* Menlo Park: Benjamin/ Cummings.

Vermeij, G. J. (1993) *A Natural History of Shells.* Princeton, NJ: Princeton University Press.

Vollrath, F. (1988) 'Untangling the spider's web'. *Trends in Ecology and Evolution,* 3, 331–5.

Vollrath, F. (1992) 'Analysis and interpretation of orb spider exploration and web-building behavior'. *Advances in the Study of Behavior,* 21, 147–99.

Vollrath, F. (1992) 'Spider webs and silks'. *Scientific American,* 266, 70–76.

Watson, J. D., Hopkins, N. H., Roberts, J. W., Steitz, J. A., and Weiner, A. M. (1987) *Molecular Biology of the Gene* (4th edn). Menlo Park: Benjamin/Cummings.

Weiner, J. (1994) *The Beak of the Finch.* London: Jonathan Cape.

Williams, G. C. (1992) *Natural Selection: Domains, Levels and Challenges.* Oxford: Oxford University Press.

Wilson, E. O. (1971) *The Insect Societies.* Cambridge, Mass.: The Belknap Press of Harvard University Press.

Wolpert, L. (1991) *The Triumph of the Embryo.* Oxford: Oxford University Press.

Wright, S. (1932) 'The roles of mutation, inbreeding, crossbreeding and selection in evolution'. *Proceedings 6th International Congress of Genetics,* 1, 356–66.

*I*NDEX